Europe: I Struggle, I Overcome · Wilfried Martens

Europe: I Struggle, I Overcome

Wilfried Martens

Foreword by Angela Merkel

Wilfried Martens
Handelstraat/Rue du Commerce 10
B-1000 Brussels
Belgium
info@thinkingeurope.eu
www.thinkingeurope.eu

This is a publication of the Centre for European Studies, the official think-tank of the European People's Party (EPP). This publication receives funding from the Community and the European Parliament.

The Centre for European Studies and the European Parliament assume no responsibility for facts or opinions expressed in this publication or their subsequent use.

Original Dutch edition published by Lannoo, Tielt, 2006

ISBN 978-3-540-89288-5   e-ISBN 978-3-540-89289-2
DOI 10.1007/978-3-540-89289-2
Springer Dordrecht Heidelberg London New York

Library of Congress Control Number: 2009929707

© Centre for European Studies 2008

This work is subject to copyright. All rights are reserved, whether the whole or part of the material is concerned, specifically the rights of translation, reprinting, reuse of illustrations, recitation, broadcasting, reproduction on microfilm or in any other way, and storage in data banks. Duplication of this publication or parts thereof is permitted only under the provisions of the German Copyright Law of September 9, 1965, in its current version, and permission for use must always be obtained from the Centre for European Studies. Violations are liable to prosecution under the German Copyright Law.

The use of general descriptive names, registered names, trademarks, etc. in this publication does not imply, even in the absence of a specific statement, that such names are exempt from the relevant protective laws and regulations and therefore free for general use.

The processing of the manuscript was concluded in 2008.

Cover design: WMXDesign GmbH

Printed on acid-free paper

Springer is part of Springer Science+Business Media (www.springer.com)

O Lord support us all the day long
Until the evening comes,
The shadows lengthen and the busy world is hushed,
The fever of life is over and our work is done.

Then Lord in your Mercy,
Give us a safe lodging,
A Holy rest and peace at the last.
Amen.

        John Henry Cardinal Newman

        (*The Oxford Book of Prayer,* 1988, p. 101)

Foreword

Wilfried Martens is one of the most distinguished politicians to come from Belgium in the last fifty years. In his long political career, he has always fought for the implementation of his goals with passion and deep conviction.

What has to be mentioned first in this context is his successful advocacy of federalism in his home country, Belgium. Already in his time as chair of the CVP youth league from 1967 to 1972, groundbreaking manifestos on Belgium's federalisation were written under his auspices. Later on, as President of the CVP, he succeeded in negotiating the Egmont Pact, whose essential elements form part of today's federal Belgian Constitution. He served as Prime Minister for more than a decade, and in this role he convinced the regions and linguistic groups of his country, which were often at odds with each other, of the necessity for cooperation and solidarity.

From early on Wilfried Martens has been a staunch European. The work of the great European statesmen Konrad Adenauer, Robert Schuman and Alcide de Gasperi was an early source of inspiration for him and continues to stimulate him until this very day. As an adviser to Leo Tindemans, Wilfried Martens dealt with fundamental questions of European integration already in the 1960s. His contributions to Europe make up a long list. Particularly important was his work for European and international Christian Democracy: he was a founding member of the European People's Party (EPP), President of its Programme Commission from 1976 to 1977, and he has been President of the EPP since 1990. He has also chaired the Christian Democrat International (CDI), the European Union of Christian Democrats (EUCD) and the

EPP Group in the European Parliament for several years each. Without his efforts, the EPP would not be Europe's strongest political force today.

Wilfried Martens' unflagging commitment as a Christian Democrat, as a Fleming and a Belgian, as well as a staunch European, is rooted in his constant effort to live out Christian values and translate them into politics. He himself concludes that his whole life and his entire political career actually consist in the attempt to bring people together and reconcile them in the spirit of Christian Democracy.

Although Wilfried Martens modestly asks readers to judge for themselves whether he has been successful in this, I would like to pre-empt them – and him – and say already at this point: Yes, dear Wilfried, you have succeeded, and for this you deserve our gratitude.

Dr. Angela Merkel,
Chancellor of Germany, President of the CDU

Acknowledgements

The book at hand is an updated version of my autobiography, with an emphasis on my engagement in European integration. In addition to new content, there is much that has been excerpted and translated from earlier versions of my autobiography, namely *De memoires: Luctor et emergo* (Tielt, 2006) and *Mémoires pour mon pays* (Bruxelles, 2006). I would like to thank the publishers, Lannoo and Editions Racine, respectively, for their kind permission.

I wish to offer my special thanks to *Hugo De Ridder, Steven Van Hecke* and *Kostas Sasmatzoglou*, who assisted me in writing this book.

I also thank *Ivan Delibašić, Richard Ratzlaff* and *Maureen Epp* for the editing and also *Peter Flynn, Ingrid Goossens* and the late *Suzanne Walters* for the translation.

Table of Contents

Prologue 1

Chapter I **Looking Back on My Life**.................................... 5

The Nucleus of Everything................................ 5
The Patience of Job 7
Children: The Happy Centre of Life 8
A Flemish Activist 9
The State of the Union................................... 11
"Give Us Weapons!"...................................... 15
King Baudouin 16
A Convinced Federalist.................................. 18
Life Is a Big Mystery..................................... 21

Chapter II **Conviction and Responsibility** 23

A New Generation 24
Promoting Innovative Ideas............................. 25
Intellectual Roots.. 28
Leader of the Flemish Christian Democrats 29
"We Can Win Again!" 30
Respect for Life.. 31
Leo Tindemans.. 33
Pioneering Work with Lücker........................... 35
Vision and Initiative 37
A European People's Party 38
Christian Democratic Principles........................ 40
Belgium or Europe? 41

Chapter III — In Charge of a Restless State 43

- 16, rue de la Loi 45
- A Referee and a Seeker of Consensus 45
- Deciding by Consensus 47
- Unionist Federalism 48
- Towards Institutional Adventurism? 49
- A Global Concept 50
- The Civil Rights of a Federal System 51
- My Heart Is Almost Torn, Writing These Words 51
- Give Me the Liberals Any Day 53
- Devaluation 54
- A Battle for the Future 58
- No Turnaround 59
- Guy Verhofstadt 60
- The King Refuses to Sign 62
- A Difficult Quest 64
- With Beating Heart 65

Chapter IV — Small Country, Large Responsibilities 67

- Rekindling the Cold War 67
- Ethical Considerations 69
- Local and Socialist Opposition 71
- Visit to Washington 73
- Allies… 75
- The KGB Listens In 76
- No More Waiting 77
- Mobutu 78
- Old Debts 79
- Not a Friend 80
- Lubumbashi 81
- Rwanda 81
- Bush Warriors 82
- The Peace Mission to the Great Lakes 83
- A Tragic President 85
- Pilgrimage Through Africa 86

Chapter V — My Road Towards Maastricht 89

- The Giscard–Schmidt Tandem 89
- First Time in Maastricht 90
- My First Presidency (1982) 91
- Towards a Re-Launch 92
- My Second Presidency (1987) 93
- On a Pilgrimage for Europe 94

Ambitions for a Top Job 95
The Long Way to the Euro 96
The European Trio 97
... Versus Mrs Thatcher. 98
The Fall of the Wall 99
German Reunification 100
Towards a Political Union 102
Twice in Rome 103
A Role for the EPP 104
Among Christian Democrats 105
My Battle for Brussels 106
Mid-Term Review. 108

Chapter VI   **"O Hellenic Shore Where Our Fathers Dwell ..."** *109*

Rekindling an Old Flame 110
My First Congress 110
Celtic Warriors 111
Deepening and Widening 112
Opting for Aznar 113
No Reservations 115
Stalwarts 117
Back to the Barricades. 118
The Tories 119
A Party Matter 120
Fraktionsgemeinschaft: Yes or No 121
Strong Foundations 122
Hellenic Shore 124
Unity in Diversity. 125
Defeat at Corfu 126

Chapter VII   **The Fall of Santer.** *129*

My Election as EPP Group Chair 130
The Santer Commission 131
"Neutral" Enlargements 133
Hostage in Porto. 135
A Role for the Parliament and Another for the EPP 137
Our Italian Mainstay 139
Forza Italia After All 141
The "Cottage" Talks on the Rhine 142
A Clear Majority. 144
Doubts About the Opposition 145
The Rift with Prodi 145
The Athens Group 147

François Bayrou and the EPP Council 147
The Leading Body Within Parliament 148
Dialogue with the Orthodox Churches 149
Elected Once Again 151
The Lost Battle for Santer 152
Prodi as New President of the Commission 153

Chapter VIII   **A *Cordon Sanitaire* Around Austria** *155*

Friendship and Solitude 156
My Heritage 157
From Bad to Worse 159
Déjà-vu 160
Finally the French 161
Exit the EDU 163
A *Cordon Sanitaire* Around Austria 164
Keeping a Low Profile 165
An Example and a Pioneer 166
Disaster at Nice 167
Fireworks at the EPP Congress 168
Verständnis mit Schäuble 169
Brok's Energetic Group 169
Constitutional Ambitions 170

Chapter IX   **"He Did Not Take the Fruit of Those Who Had Planted the Tree ..."** *173*

In the Driver's Seat 174
Tête-à-tête in Lisbon 175
Right up to the Tarmac 175
Three-Cushion Billiards 176
Among Believers 177
A Changed Field of Influence 178
A Strong Position Confirmed 179
Primarily National Ambitions 180
The Hard Core 181
The Party of Continuity 183
Political Leadership 185
A Cultural Homeland 187

Chapter X   **Pushing Forward in Europe and Beyond** *189*

The EUCD's Final Role 189
The Hungarian Struggle 191
Fidesz – Hungarian Civic Union 192

The Velvet Revolution 193
… and Velvet Separation. 195
The Beginning of the End of Communism 196
Baltic Strength 198
The Mediterranean Island States. 200
From a Social Democratic to a People's Party 201
Hope for Change and the Rule of Law 204
To Be Right and to Be Proven Right 205
The Call of Bosnia 208
"Across the Country of the Illyrians" 209
The European Choice of the People 210
In Search of an Identity. 212
The Breadbasket of Europe. 213
The Caucasian Conundrum 215
Test Case Turkey. 217
No Muslim Democrats 218
Together and With Us 218
Moving to the World Stage 220

Epilogue *223*
Bibliography *233*
Index of Names *237*
Annex. *243*
Notes *251*
Photos *253*

Prologue

I was born on a small farm in Flanders four years before the outbreak of the Second World War. There was nothing special about my background to suggest that I would ever play a role of any significance. But historians have shown that seminal movements often have their origins among the peasantry. Indeed, it is said that my country has long been ruled by people from the countryside. If that is true, then I am one of them. I grew up in a poor, rather old-fashioned world. The history of my family is the story of those 'little people' so typical of Flanders. My forebears worked as farm labourers or in cottage industries on piece wages, living a life of poverty. An exception to this was my oldest known ancestor, Jacob Martens, who was a bailiff in his commune from 1602 to 1618. But most of my ancestors belonged to the working class, had many children and lived in anonymity.

The fact that I became a politician nonetheless and devoted my life to public service can be attributed to the philosophers and politicians who, after the war, laid the foundations for the European society of my youth. My own political awakening was the result of their influence. Philosophers like Jacques Maritain, Emmanuel Mounier and Paul Ricœur gave the West a new intellectual climate. They established the foundations of Europe as we now know it. This new intellectual climate brought forth politicians of exceptional stature, individuals marked by a war that had transformed Western Europe into a heap of rubble. They included Konrad Adenauer, Robert Schuman, Alcide de Gasperi and Jean Monnet. These philosophers and politicians remained a constant source of inspiration for me during my entire time in politics.

The philosophy of Jacques Maritain, for instance, centres on the human person. Humans should, first and foremost, become more human(e). The process of humanisation unfolds, for Maritain, according to the Christian vision, as opposed to the atheistic humanism of socialism/communism and fascism. It is only through a synthesis of two distinct yet never entirely separable orders that one can achieve an 'integral' humanism. Emmanuel Mounier believed there would be a renaissance, a rebirth of person and community. He strove not for the emergence of a new type of person (unlike in fascist and communist ideologies), but for a restoration of the absolute value of the human person. Like Maritain's, his thinking is Christian, but he leaves room for believers of other faiths as well as for unbelievers. Mounier was an inspiration for and friend of Paul Ricœur (1913–2005), who in turn strongly influenced my political engagement. This primarily French strand of personalism had its German counterpart in the work of Max Scheler and Romano Guardini.

Christian Democratic parties, which after the Second World War replaced the confessional Catholic and Protestant parties that were strongly influenced by the churches, took their inspiration from this philosophy of personalism. Various elements of Maritain's and Mounier's philosophy constitute an integral part of Christian Democracy: the integration of the spiritual or transcendental into its vision of humanity; the role of religion, and of Christianity in particular, as the final end or goal of existence; the primacy of community over the subsidiary, decentralised state; societal pluralism; the personal and social significance of labour and private property; and so on. Moreover, Christian Democracy shares the criticisms of Western rationalistic culture, materialism and unbridled capitalism expressed by the personalist thinkers.

Luctor et emergo

My aim in this book is to record how I put this inspiration into political practice: from my early days as a student leader to my time as a Flemish radical, then as President of the Christian Democratic youth movement, party President of the Flemish Christian Democrats, Prime Minister of Belgium, President of the European People's Party and chair of its group in the European Parliament.

After twenty five years of growing prosperity and political equilibrium, even Belgium, my own country, has been going through a serious crisis now for eighteen months. In this book I want to show how complex, but also how inspiring it is to lead a country that intersects Latin and Germanic cultures. Our two Christian Democratic parties have also always been at the forefront in Belgium in finding solutions that were acceptable to both French and Flemish speakers. Unfortunately, during a hard and painful period in opposition (1999–2007) the two parties have grown apart. Among many people this has provoked a nationalistic response. The futility of this has been demonstrated in recent months for the whole world to see. It has had a profound effect on me. But I am striving to the best of my ability to bring about a new rapprochement between the Christian Democrats of the north and the south. If this were to succeed, the Belgian impasse would immediately be resolved.

I hesitated to write these memoirs. It has been an exciting but all-absorbing pursuit. Every life story is a series of successes and failures. We like to remember the successes. We tend to shut out failures. Yet the two are closely intertwined. A failure sharpens the desire for battle and many a victory has been born of setbacks. *Luctor et emergo*; I struggle and I overcome. Incidentally, this is the motto of the province of Zeeuws Vlaanderen (Zealandic Flanders), not far from my native village Sleidinge. I struggle and if necessary I go against the flow. But if political life is completely governed by pragmatic concerns and *managed*, if indifference and cynicism gain the upper hand and human solidarity disappears, then personal courage is needed in order to continue to stand firm. It takes a great deal of patience and trust to endure everything, to believe in everything, to hope for everything and to persevere in everything.

I have read too many memoirs not to realise that it is a dangerous genre. The writer can choose to dredge up memories and anecdotes, as gifted speakers sometimes do in after-dinner speeches, with the principal intention of charming the audience. Some memoirs are a kind of self-glorification; others an attempt to have the last word. While some seem to have the novel as their source of inspiration, others attach everlasting value to diary notes. My own goal in this book is to give politically aware citizens in Europe – the 'value-added seekers' – a personal testimony and to share my experiences of more than half a century of political life at both the national and European levels.

Is this ambition aiming too high? After all, the brain works very selectively and usually becomes more feeble as the years go by. Even though I have forgotten some things, my memory of the past is crystal clear when I am confronted with parliamentary archives or tape recordings of speeches and interviews. I then relive these events, as it were: I remember the feelings I had as I stood there on the podium and all kinds of details come into my mind, and I am taken back to the events of that period. For this intensive work I have had the assistance of a long-time observer as well as young academics who have done their theses on areas of my political activity.

This teamwork of youthful knowledge and elderly experience has many times led to heated discussions, with the principal concern being to keep as close as possible to the historical truth, insofar as it actually exists. The writing of my memoirs has in this respect been an exciting journey through my own life. By reviewing documents, pictures and sound bites, I was continually confronted with my pronouncements and opinions from years ago. Even though they may have evolved since, I have nevertheless taken care to reproduce my words and deeds from those days as accurately and as truthfully as possible.

My entire life, my entire political career, has consisted in bringing people together or reconciling them in the spirit of Christian Democracy. I leave it to the reader to decide from my life story whether I have succeeded in this. As far as I am concerned, I can testify that I always acted according to my conscience, persistently and with a deep faith.

I dedicate these memoirs to my children Chris, Anne, Sarah, Sophie and Simon, and to my grandsons, Alexander and Lucas, as well as to all those who strive to know what political commitment can mean in the life of a man.

Brussels, 31 December 2008

Chapter I

Looking Back on My Life

I am now seventy two, and when I look back on my life, images of my youth appear very vividly before my eyes. Like many others I have come to realise how much of my life was formed during those years. My inquiring mind gathered the seeds, which later grew to maturity in my temperament and character.

I am a farmer's son. Four years before the outbreak of the Second World War, I was born in a hamlet near the country village of Sleidinge, twelve kilometers north-west of Ghent. The countryside there is flat and there are a lot of canals. The Netherlands and Zeeuws-Vlaanderen (Zealandic Flanders) are just up the road. The Westerschelde River is a five-hour walk away. Our house was halfway between Ghent and Eeklo in the middle of a triangle formed by the Leie Canal, the Ghent–Terneuzen Canal and the Leopold Canal, and I can recall vividly the country lanes from our farmhouse to the village, the colourful corn and maize fields in the summer, the white painting-like landscapes in the winter.

The Nucleus of Everything

All of us carry a little bit of our youth with us into adulthood. *You can take the boy off the farm, but you can never take the farm out of the boy.* Many of us idealise our youth, both its good and bad moments. The long walks on wooden clogs, twice a day, to the village school; rye bread and buttermilk porridge; the assault by German artillery in May 1940 and the horses gone mad, running in all directions away from the farm; stealing salted pork towards the end of the war.... These became the stories I would later tell to inquisitive journalists.

What I mainly remember is how hard life was on the farm. My father would rise at four and drive his horse and cart to the neighbouring farms to collect containers full of milk. With much effort and quite a bit of noise he unloaded the heavy containers at the dairy near our house. This noise was our alarm clock: time to get up. As a young boy I had to help feed the cows and the pigs every day, and afterwards I headed off on foot or by bike to Mass in the village. After school and on holidays too we had to help make the mash for the animals or churn butter. We always had to go to bed early. Even during the summer, bedtime was at seven or eight in the evening. And every evening we would pray the Rosary.

All this sounds very much like a Flemish country novel. We lived our lives against the backdrop of this great synthesis, the Catholic faith. Religious belief, that great certitude, formed the nucleus of everything. Things were fine that way. Of course, my parents probably had their doubts and questions from time to time, but they had the moral strength to struggle on and even do something quite revolutionary. Despite their lack of means they gave all their children the opportunity to study. That was quite an achievement at the time. My parents also knew their own shortcomings and failings; they were not destined to be canonised as saints. But they were pious and deeply religious and lived their lives according to the strict moral principles of the Roman Catholic Church, strict to the extreme and without the slightest exception to the rule.

My father died during the war and my stepfather died only ten years later. So my mother, Virginia Estella, looked after the education of her five sons, of whom I was the eldest. My mother was a strong, active and extremely gifted woman. Unfortunately, she never had the opportunity to continue her education. Few women were allowed to do so in those days. She raised a large family all by herself: she cooked, made all the clothes for the children and also helped in the fields. Like all her sisters, my mother had worked as a young girl as a servant in the villas of wealthy families on the outskirts of Ghent. This was the same world as the one described by the Hungarian author Sándor Márai. In his *Confessions of a Haut-Bourgeois* (1934), he writes that "the servants slept in the kitchen, which was so vast and spacious it was almost like a flat. Some family houses in the countryside had as many as ten or twelve rooms. The cook and the servant had to sleep where they cooked and did the dishes" (Márai, French version, 1993, p. 64, our translation).

Being a farmer was not highly regarded those days. Most farmers were so poor they could not even afford to buy their own piece of land, although this was the most fervent desire of many. The usual practise was to rent a simple farm. My parents' main goal in life was to buy their own farmhouse and yard. But I witnessed the deep disappointment experienced by both my father and my stepfather, who despite all their years of hard work were never able to afford to buy land.

There was little room for tenderness and affection in our farming family. I once said that my inner life was like that of a shy deer that appears only to disappear again as suddenly and as quickly into the undergrowth. My life would have been very different had I had a sister or had there been a girl in the neighbourhood. For years, I cherished an ideal image of Woman, inspired by my beautiful mother. I have the greatest admiration for mothers of large families. I am always touched when I meet them. But in the course of time I had to change my image of women. I discovered very late the world of girls and young women. I would have become a different man had I grown up in the company of them. Gradually, I discovered that women also have shortcomings; like men, they can be lazy and messy and even untrustworthy. I met women who were as flighty as butterflies and who were very open sexually but, at the same time, were incapable of any form of tenderness.

The sexual revolution erupted in 1968. That was a time when the women's movement made great strides forward. Those were also the halcyon days of the Christian Democrat youth movement, to which I belonged; both movements left a deep impression on me and my peers. It was during these years that the stage was set for a broad, dynamic movement in which women would play an extremely active role. They fought for sexual liberation and equal rights for women and men alike. The Pill and other forms of contraception freed women from the immediate perils of intercourse, one of the great risks of the past. Making love became freer and happier. Both partners became more self-aware and equal. Many women were better off as a result.

The Patience of Job

I was married in 1962, following my studies at university. My wife and I had two children: a son, Chris, and a daughter, Anne. In 1978 our son, who was then twelve-years-old, had to have his leg amputated following a serious car accident. It took me years to come to terms with my son's severe handicap, but there was always the hope that he could live a fairly normal life. I thought at the time of the amputation that we had experienced the worst, but that was not the case.

When Chris turned twenty one he had to be treated for mental illness. It was the beginning of a long period of agony. Those who have never experienced such agony at close hand do not realise how great a trial and a tragedy it is for parents, who continually ask themselves if they are to blame. It is said that you need the energy of Prometheus and the patience of Job to live with such an illness.

My son's illness had a profound effect on my work as a politician. Initially, it was not clear what was wrong, but after a year or two it did become clear that the process was irreversible. The illness then began to weigh considerably on my work as Prime Minister. I was torn apart inside, having accepted this heavy task, but also wanting to support my son.

The wound grew deeper with the years and had serious consequences. I became less self-confident, less decisive and more vulnerable. As a young man I had never hesitated for a moment. Now I became more cautious; I began to reconsider, and reconsider again. It even went so far that I could not sleep at night. This intense pressure slowly began to recede when, in 1992, I decided to fully dedicate myself to the presidency of the European People's Party, following a period of twelve years as Prime Minister. Slowly, I started to learn how to live with my son's illness, realising that it would never heal or go away.

However, a second tragedy was awaiting my family. Following a long and painful separation from my wife, I went to live alone in Saint-Gilles in Brussels. This was far from easy for me, as I had to overcome an enormous moral obstacle, since I came from a very Catholic background and held extremely orthodox views on marriage. It took many years before I could overcome that moral obstacle and achieve some degree of balance.

Children: The Happy Centre of Life

I got to know my new partner in 1988. That we had children together made things turn out in an even more positive sense. I was sixty when our twins Sarah and Sophie were born and sixty four when I became father to our son Simon. The children have given my life a totally different, new and positive purpose. Thanks to them, I did not fall into a deep depression.

No matter how terrible a divorce is, I had the courage to take that step. Catholics have often been accused of being hypocrites: "They do not really divorce; they do not remarry, they just have a girlfriend". Divorce seems to be an everyday occurrence nowadays. That was far from being the case some years ago. I was struck by something Mazarine Pingeot, François Mitterand's natural daughter, once said in an interview regarding her father: "I can only guess why he never divorced. I never spoke to him about it myself. In the 1970s divorce was still very damaging to a politician's reputation, but I do not think that bothered him much".

These profound changes in my private life were not something I spoke about in public. When the twins were born, I did not feel the slightest urge to spread the news. Having children when you are sixty is life changing. You think about it a lot or, rather, you are aware that you have brought them into a world that is full of risks. I hesitated for a moment but then I found the courage to go ahead with it.

Marriage has the finality of being permanent; according to the Roman Catholic Church, it is an unbreakable sacrament. But there are situations in which a marriage ceases to exist. The other churches have recognised this. My divorce had no effect at all on my religious convictions. It has nothing to do with whether I believe in a personal God, which I certainly do. The Gospels are the source of inspiration for my political action and I read the Bible in the same way I would an exciting novel.

I do not wish to mount an apology for divorce, on the contrary. A divorce is a heavy ordeal. It is like being shipwrecked. Both partners most probably share the responsibility for the failure. The children, especially young children, are the chief victims. Despite everything, it still remains a valuable thing to recognise the truth and to gather the courage to start again instead of continuing to live an illusion.

In the past, I used to fit the long-familiar picture of the working father. It did happen that I was so focused on my work as a politician that I did not hear what my children were saying to me. I used to try to spend time with them on weekends, when we often went to the seaside together. I agree completely with the German journalist, Sandra Kegel, who wrote that it is high time that we supported the younger generations and freed them from the pretence that they have to be perfect in their lives and in their families. Those who have children simply cannot do everything perfectly.

Children bring with them chaos, sleepless nights and total unpredictability. Children create limits for their parents but at the same time they open up horizons. Children do not only limit their parents; they focus their attention on what really matters. Perhaps they teach us to have that little spark of courage to improvise, a vital bit of fantasy that makes the adventure of family life worthwhile. Otherwise, children would no longer be at the happy centre of life – and humans would merely be living in a reservation designed to keep the species alive.

I no longer run off to meetings in the evening, and this I feel as a great liberation. I did so for thirty or fourty years. I sometimes ask myself in retrospect what my life consisted of. I hardly had to the time to read a book then. I felt an emptiness that continued to grow, but happily I was to fill that void later on. When I meet my former colleagues from the European Parliament and see them at work, away from home week after week and always off again on some trip, I often ask if they find any time at all to be with their children. Children form the heart and the future of society. Without children there is no hope. And He said to them, "Whoever receives this child in My name receives Me, and whoever receives Me receives Him who sent Me; for the one who is least among all of you, this is the one who is great" (Luke 9:48).

A Flemish Activist

These memories and reflections come to me when I look back, as I do now, on my youth and on my parents, and especially on my mother. To a very large extent, my upbringing shaped my attitude towards women and fatherhood. My family and private life would have been completely different had I grown up in the city. My political commitment is also deeply rooted in the place I was born.

Sleidinge, my village, was very much a closed community. Each village spoke its own dialect, which could hardly be understood by people a few villages away; this was the case for all villages in Flanders. Our use of dialect was one of the many reasons why French speakers looked down on us with some disdain. The medium of conversation between the mayor, the notary, the parish priest and the factory owner was invariably French, both in the countryside and in the towns of Flanders. This was not surprising: because of the lack of Dutch-language universities, intellectuals were obliged to obtain their academic qualifications in French until as late as the 1930s.

From my fourteenth year, I abandoned my dialect and, along with about five fellow students, began to learn standard "educated" Dutch (Algemeen Beschaafd Nederlands or ABN; literally, "standard civilised" or "educated" Dutch). The adjective "beschaafd" was later dropped and the term is now Algemeen Nederlands, Standard Dutch. Our primary example was the language spoken on the radio. Many of the secondary school teachers of the day spoke a sort of "in-between" variety and certainly did not encourage the use of ABN or Standard Dutch. On the contrary, they considered it an expression of Flemish extremism. The struggle for a pure language really excited me. It lay at the root of my political awakening.

This step towards Standard Dutch was probably one of the most fundamental decisions of my youth, both for me and also for many other young people. It was a daily battle. It demanded strength of character; you had to arm yourself constantly against exclusion. Choosing resolutely for Standard Dutch was proof that you had higher ambitions, particularly if you came from a less well-off family. Anyone who articulated every word also cherished plans to leave his or her background behind and move on. This led to alienation and even to a break with family. Initially, speakers of ABN were teased by their friends, brothers and sisters. They imitated you and if they noticed that it had no effect they left you alone.

Speakers of ABN therefore sought the company of fellow speakers. In this way a language elite was formed. You felt comfortable only when you could speak your language in another province where you would be praised for your perfect pronunciation. Those who promoted the language argued that Flemish people should speak Standard Dutch; otherwise, they would remain an inferior people and would never be able to compete with French-speaking Belgians who did speak their language well. The ABN speaker made a constant effort to refine his or her language use, and that even meant speaking like people from Holland, which irritated many. "They have gotten too big for their boots", people would say. Do not forget we were very young students at the time and, paradoxically enough, we had to distance ourselves from our own people in order to educate them.

Through ABN circles I came into contact with many pro-Flemish people. I got to know the basic elements of what was then known as the *Vlaamse kwestie* or the "Flemish question". Despite the fact that the Flemish were in the majority in Belgium, the French speakers called the tune. This took the form of an overrepresentation of French speakers in the government, the administration, the army, the law courts and the diplomatic corps. Those in industry and in the arts also looked down with pity on anyone who insisted on speaking Dutch. French was actively promoted in and around the capital, Brussels, through social pressure and also through so-called language censuses, as a result of which more and more boroughs acquired bilingual status. We called it "the Brussels oil spill": French-speaking people moving into the Flemish suburbs of Brussels.

Some people – and there were quite a few prominent ones among them – thought that the problem would resolve itself, that the majority on the Flemish side would manifest itself in time both in the economy and in politics. Personally, I thought that our goals were unattainable by this approach, and that it might even be dangerous. I was totally convinced that the best solution was to set up a federal state with considerable autonomy both for Flanders and for Wallonia.

The State of the Union

As a student at the Catholic University of Louvain, I soon became the leader of pro-Flemish student associations that wished to exert pressure on the government though protest marches and manifestos. The high point of all this was our struggle for a Flemish day at the 1958 World's Fair in Brussels, which landed me on the front page of the newspapers for the first time. The Fair was the first post-war exhibition of considerable standing. The whole world was going to visit Brussels, stare in awe at the Atomium and all the other technological advances on display. But the preparation was completely in the hands of French speakers. There was a clear risk that the millions of visitors would get the wrong impression of my country. As far as we were concerned, the Fair had to make clear that we no longer lived in the Belgium of 1830 but that the country comprised two communities, each with its own character, language and culture, and both linked by a common tradition and a will to live together.

As a result of the pressure exerted by our demonstrations, the memorable Flemish Fair Day was held on Sunday, 6 July 1958. The rectors of the Flemish universities, ministers and provincial governors competed with each other for a front-row seat. During the ceremony, a small group of "Flamingants" (Flemish nationalist extremists) led by a Flemish militant, headed for the French pavilion, planning to paint over the French signs. The incident was covered in the international press, especially because the French President, René Coty, was to arrive on an official visit two days later.

For a long time, and even later in Parliament, I was accused of having thrown those paint bombs at the French pavilion, hence showing my disdain for French culture. But I had a watertight alibi: that very day I was studying for my exams in Louvain and following everything on the radio. I was not even present at the Flemish Day I had helped bring about.

At the following Yser pilgrimage, an annual day of remembrance for the Flemish who perished on the battlefields of the First World War, I was allowed as a twenty two-year-old student to address the crowd with a Flemish "state of the union" speech, as it were. I delivered the speech then with all the sharpness of my youth.

> Those in power are using the World's Fair as a pretext to propagate in the most extreme way possible French-speaking centralist forces in Brussels. We now know that Brussels will become the capital of Europe. Those in power will also use this to keep the Flemish as far removed as possible from crucial areas of power. And in truth it will be easy for them: the Flemish people do not even have the apparatus they need to assert their culture and community within the Belgian state, never mind the European institutions! As long as we do not possess our own institutions, we will be doomed to remain inferior and will ultimately disappear as a people. But once the Flemish achieve autonomy and are able to provide for the specific needs of the Flemish economy and are capable of pursuing a progressive social policy in order to achieve prosperity for the whole of the Flemish community, and promote and encourage the Dutch language and culture from above, then we will be restored as a people. Then and only then will we able to work to build a European community without fear for our own existence.
>
> Young Flemings! We will be the first Dutch-speaking generation to be taken up into a wider Europe. And just like other European youth, we must become highly educated, complete in every human sense and convinced representatives of our culture. But unlike the youth of other countries, we have a huge deficit to overcome. Dutch, that refined and educated spoken Dutch, is something we have not yet mastered, and we often lack the courage to use these pure spoken words in our everyday Flemish lives. In addition to this education, we also have to make a concerted effort to ensure that the struggle for Flemish emancipation succeeds. We know the consequences of this struggle: persevering in our study, sacrificing our security, severe asceticism. We have expressed it differently here but it is also the motto of the Yser Pilgrimage: "All for Flanders, Flanders for Christ". Our work for Flanders is made less burdensome by the knowledge that we have found unity and that we can work together with youth of all opinions.

Though this speech may sound romantic and flowery now, it received a lot of attention at the time. I was elected president – as the sole candidate – of the umbrella student association for Flanders for the academic year 1958-1959. The following year, I became president of the student union at the Catholic University of Louvain. My regular contacts with Flemish Christian Democrats in Parliament, who were expressly favourable to the Flemish cause, also date from that period.

Following my studies in Louvain, I was called to the Bar in Ghent in October 1960. I was then twenty four-years-old. During my final year as a student I was elected to the leadership of the Vlaamse Volksbeweging (Flemish People's Movement), an association of independent members whose cherished ambition was to act as a pluralist pressure group for Flanders. It was then that I studied federalism, my most important source of inspiration being *Études sur le fédéralisme* by Robert R. Bowie and Carl J. Friedrich (1960). The European Commission had commissioned the work, which comprised a study and comparison of various federal systems, including those of the Federal Republic of Germany, Austria, Switzerland and the United States. *Études sur le fédéralisme* helped me focus my ideas in my search for reasonable and responsible systems in order to reform the Belgian state according to the principle of "autonomy where possible, strong central power where needed".

I soon put my ideas for a federal Belgium down on paper and presented them at the congress held by the Flemish People's Movement in Antwerp on 4 February 1962. Through the years, many have pointed to this speech as a feasible sketch of how to move from a strictly unitary state to unionist federalism. For thirty years, this notion has formed the golden thread in my political struggle as a party leader and as Prime Minister. The ultimate reward arrived only in 1993, when federalism was enshrined definitively within the Belgian Constitution. Briefly, this is what I said at the time:

> The political events of the past year show that the reforms carried out to the structure of the Belgian state have become of primary importance for the country, given that the partial measures taken so far have proved to be insufficient and have only given rise to confusion. The Flemish and the Walloon communities have to acquire effective autonomy in a federal state system within their common fatherland.
>
> Firstly, the territories of the Flemish and Walloon communities have to be separated by a language border and recognised as separate homogeneous cultural areas. The territory will form the infrastructure of the federal states of Flanders and Wallonia, as it were. If the Flemish People's Movement has been conducting a campaign for the maintenance of something that has been recognised for centuries as Flemish soil, it has not been doing so because it considers that soil as something holy but rather because that soil is needed as a foundation upon which the Flemish community can allow its own institutions to function.
>
> The undermining of the cultural homogeneity of Flanders has to be resolutely stopped, which is why the language border has to be drawn. In a federal system, the language border will function as the border between the members of the federation. The language border will be written into the Constitution and will be part of the fundamental agreements upon which the federal state will be built.

Secondly, the Flemish and the Walloon communities should acquire their own legally recognised assemblies, governments and judiciary.

In this way, the will of the Flemish people will be heard for the very first time in such an assembly, following general and local elections based on absolute proportional representation. Each assembly will hold sovereignty over the territory it is empowered to govern and the sovereignty of each member of the federal state will be enshrined in the Constitution. The federal government shall see to it that each separate government carries out its policies according to the law.

Furthermore, the institutions of each member need to be complemented by the establishment of a social and economic council which will exercise advisory and possibly mandatory authority, whose purpose will be to involve workers, management and the self-employed more closely in economic policy.

Thirdly, the Flemish and the Walloon communities should each acquire a proportionate degree of political autonomy. The Flemish People's Movement does not foresee complete political autonomy for either Flanders or Wallonia. This would only undermine the very foundations of the federal system, which is essentially a system designed to distribute political power between the federation and each of its members. Each member should have complete cultural autonomy, that is, its own policy and legislation under effective supervision, which may or may not stem from a vote of agreement on the budget. This cultural autonomy will comprise education, in complete accordance with the language laws, culture (in all its forms), tourism, physical education and sports, radio and television. This political power is essential. In this respect, Flanders has to overcome a deficit in higher and further education and also with regard to a more rapid completion of the process of democratisation. It urgently has to change its educational programmes, which till now have been moulded according to the French mindset and which have not been adapted to the specific nature of Flemish youth.

This same necessity, which is also informed by the need to develop a separate policy on population and housing, for example, also applies to competence in public health and family life. As far as internal affairs are concerned, such matters as the running of borough and provincial councils in their totality, the structure of constituencies, borough financing and the application of language laws in all areas of governance are essentially all aspects of independent Flemish and Walloon political power.

Finally, the members should avail themselves of the broadest possible competence in matters of regional socio-economic policy, agriculture and public works.

"Give Us Weapons!"

On 8 November 1962, a few months after this speech, the first bill on language was passed in Parliament, which fixed the language border definitively. A year later a second proposal, which settled matters of language use for the capital, Brussels, was placed on the political agenda. The paper proposed that Brussels Capital Region should become officially bilingual. The region comprised the nineteen Brussels boroughs and six other Flemish boroughs bordering on the capital that would become bilingual from then on: Drogenbos, Kraainem, Linkebeek, Sint-Genesius-Rode, Wemmel and Wezembeek-Oppem. This compromise proposal was rejected, however, by the Flemish Group of the Christian Democrats, placing Prime Minister Theo Lefèvre's government in serious difficulty.

It was in one of these contested boroughs that I delivered a true war speech:

> What we are witnessing here is a poker game by property developers. They want to sell the land to the highest bidder and that bidder is a French-speaking capitalist. But there is more at stake than that. The government wants to set up schools and services in Louvain and in the other Flemish cities that house our national institutions. We will have to go on the offensive and demand that all companies in Flanders – no matter how small they are – become Dutch-speaking. We reject out of hand any special allowances for French speakers in the Flemish boroughs. The Flemish members of Parliament should also reject the Lefèvre government's policy and the Flemish ministers should resign.

I gave this speech on a rostrum in the open air to a crowd of pro-Flemish protesters 8,000 strong. The protesters immediately began to chant, "Resign, resign". I continued: "If the conspiracy against Flanders does manage to succeed then we will let the Walloons and those in Brussels know that there will be a final march on Brussels, and its aim will be to overthrow unitary capitalist Belgium. We will organise that demonstration at an embarrassing moment for the parliament". There was a resounding applause. "If your dictate should become law then, Theo (Lefévre), there will be a revolution!"

While the protesters were cheering me on, a group of Flemish militants off to the side of the rostrum began to chant, "Give us weapons!" Though the slogan never came from my mouth I have been accused by many for a long time of having said it, even by those in Parliament. It certainly was not a slogan the organisers approved of, because when various groups of protesters also began to chant, "Give us weapons!" after an altercation with the police, one of the organisers jumped onto the rostrum and shouted, "We have a weapon! Our weapon is more effective than stones. We have the people behind us and with the people we will cast aside each politician who betrays us!"

On 2 July, two days after the demonstration, Prime Minister Theo Lefèvre handed in his resignation, which was not accepted by the King, however. This was followed by the famous conclave of prominent members of the Christian Democrats and Socialists at Val Duchesse castle. On 5 July a new compromise was reached at the historic castle. From then on, the six Flemish boroughs would form a separate administrative entity with its own constituency commissioner answerable to a new – yet to be appointed – vice-governor of Brabant, who would watch over the application of the language laws. A form of administrative bilingualism was installed in the Flemish boroughs bordering on the capital: French-language kindergarten and primary schools could be set up if at least sixteen parents should request this. No French-speaking secondary schools were allowed, however.

The new laws were passed by the Chamber and the Senate a few weeks later. Many Flemish Christian Democrat MPs said afterwards that it was with utter despair and totally against their will that they pressed the voting button on that day. It was felt to be a defeat for Flanders and provoked bitter argument and vicious reactions. I reached the conclusion then that this defeat would determine the rest of my life. During one of our debriefing sessions I stated that "action outside Parliament has proved insufficient. We can only achieve something through power, so through a large party. There is no point in our rummaging around in a nationalist opposition party. That will get us nowhere. If we were to take up the cause within the Flemish Christian Democrats, then that party would have to change. The pressure would be unbearable".

In the course of time, about five of the leading members of the Flemish People's Movement followed my advice and together we managed to gain a foothold for our federalist views within the largest party in Flanders. In fact, it would become apparent from this and other events that the defeat formed a pivotal moment in the story of my life.

King Baudouin

One of the other memories of my youth is of an event that occurred in 1950, which would continue to have an enormous influence on my life. I had been confined to bed since Easter with a serious infection of the joints. I suffered severe bouts of high fever and people even feared for my life. A professor from Ghent prescribed penicillin for me, and my godmother Emma came to give me injections three times a day and sometimes even at night. The local doctor issued this ominous judgement at the time: "Wilfried has a weak heart. His aorta valve has been damaged as a result of his illness. Wilfried can finish secondary school but after that he should only do light office work, at the very most".

Around that time, I often sat listening with bated breath to the little radio my mother had been given. On the morning of 1 August 1950, the national news carried a huge story. The royal question – should King Leopold III, who was accused of collaboration during the German occupation, return to the Belgian throne? – had taken a dramatic turn in the course of the night. In order to prevent further unrest, Leopold III had expressed the wish that Parliament should pass a law to the effect that his powers be delegated to his son, Crown Prince Baudouin. Leopold III hoped that the long-promised reconciliation would come about in the presence of the young prince.

Even though I had little education in politics at the time, I felt that this event would be of enormous consequence for the whole country. A few months previously, on 12 March 1950, 57.68% of Belgians had voted in a referendum in favour of the return of the controversial King from exile. In Flanders the figure was as high as 72% whereas in Wallonia it was only 42%. Only when Leopold III returned to his castle in Laeken on 22 July – at the crack of dawn and flanked by rows of policemen – did unrest break out in earnest, followed by many strikes, particularly in Wallonia. During a turbulent demonstration at Grâce-Berleur, the police shot three demonstrators dead. The Socialists and Communists threatened to march on Brussels.

I had been following the political events closely on the radio for weeks. "Leopold's thrown in the towel", I heard someone shout to my stepfather in a sharp voice. I tried to imagine what the young prince would feel like in such a big palace. I could not imagine then what a prominent role he would come to play in my political and personal life. On 1 August 1983, thirty three years later to the very day, I sat as Prime Minister facing King Baudouin at Laeken Castle. I had just learned from the doctors that I would have to undergo an urgent operation on the damaged valve in my heart: they planned to replace it with a new synthetic valve. The King was greatly concerned and tried to give me courage. Following a successful operation by the highly reputed heart surgeon Georges Stalpaert from Louvain, the King invited me to spend a period of convalescence at the royal residence of Opgrimbie in the woody Kempen.

Ever since my youth I had followed the work of the young Baudouin with undisguised sympathy. That feeling has remained with me all my life. My experience of King Baudouin is one of a man of great integrity, high moral standards and an extreme sense of the importance of the state. Within the Belgian system, the king is the final protector of the Constitution; he ensures continuity and also the maintenance of balance between the various institutional powers. Baudouin was King of Belgium for forty-three years. As holder of the institution, his power and influence was a lot greater than ordinary citizens might realise.

The monarchy had become so obvious an institution to the Belgians that its importance as mediator and ordering factor is often underestimated.

The monarchy is neither a quaint tradition, nor decorative, something to show on television during the Te Deum celebrations at the Cathedral of St. Michael and St. Gudula on 21 July, the national holiday. In reality, the monarchy has a pivotal position in the Constitution. It was one of my predecessors, the socialist Prime Minister Achiel Van Acker, who once said that our country needs the monarchy like it needs bread. This is completely true. There is no other head of state possible in Belgium. Moreover, the constitutional system within which the king functions is so delicate and finely honed that the limitations placed on his own political influence give natural form to his role as a referee.

A Convinced Federalist

I was the first Prime Minister younger than King Baudouin. He was objective and scrupulous in his task; for example, in solving crises in government and appointing *formateurs* – the person appointed by the king to form a cabinet, who as a rule also becomes Prime Minister – and ministers. But he was also thus in his daily life. The palace manifested itself through discrete but useful intervention. As Prime Minister, I often felt the need to make Baudouin privy to my questions, thoughts and reflections. In his company I was filled with lofty notions, such as the independence and integrity of the state, for example. It is precisely because the Constitution removes the king from political discussions and argument that he can gain the authority of distance. Moreover, he symbolises something I can only call harmony, a concern for the country in all its diversity. Of course, because of his responsibility to the government, Baudouin was extremely aware of the huge movement that was active then in the reorganisation of the state. As a gifted coworker, he himself was fully involved in the pending reform to the Belgian state. I myself am not a structuralist: history is made by people and not by programmed techniques. Working within the clear constitutional rules to which contemporary royal power is subject, Baudouin was a man of exceptional skill and sympathy for the whole of society and for all those who were committed to working for it in whatever way. In the United States this is known as compassion, and I have always seen that compassion in his speeches.

Over the course of the years, I saw Baudouin evolve from a moderate reformer to a convinced federalist. To the extent to which I could play a role, I do believe that I helped the monarchy in some way. Maintaining the status quo or stubbornly holding on to outmoded institutions would have proved harmful for the continuation of the monarchy in our country.

I often think back with nostalgia on the many hours I spent in Baudouin's company. Nowhere else can a politician be more forthright than in the king's study or sitting on the sofa at Laeken Castle. The politician knows at all times that the conversation will never go beyond those four walls. He or she feels safe there. Though some may disagree, I am totally convinced that people tell the most truthful political stories only to the king. At least that was the case in King Baudouin's day. Of course, he wrote things in his little black notebook from time to time. I often wonder whether historians will be able to glean and understand all the complexities they contain, twenty or fifty years from now. In any case, many a myth will be debunked if they are ever made public.

Baudouin usually greeted me with a firm handshake and a penetrating look. On each occasion it felt as if he was inquiring about my mood and frame of mind. Though he could be very funny at times, Baudouin was usually serious and highly conscientious to the point of being scrupulous. He never tread lightly on affairs of state. His position weighed heavily on his shoulders. He rose each morning to the not-so-pleasant prospect of being confronted for the whole day with problems of state. He had a lot of trouble trying to set all this aside.

He remarked of himself that he was useless on the telephone. He preferred to meet people face to face. But sometimes things had to be done over the phone. Much like me, he was brief, but in further conversations always gave ample explanation about the deeper significance of his approach.

King Baudouin prepared his speeches with extreme care and in great detail. He was always well informed through intensive talks with Prime Ministers and his own private secretary. I explained the government's plans to him in complete detail on a weekly basis and also informed him of counter-arguments and opposition within the cabinet or from other party members. The King constantly inquired into proceedings and sometimes gave blunt comment on matters. Next to these, some of the problems we dealt with were quite basic and practical in nature, like restoring buildings, laying safe telephone lines and even considering the royal finances.

When the weather was good we often took a stroll on the beautiful grounds of Laeken Castle. Things were more formal at the palace in Brussels. The King invited me to lunch in Laeken on a number of occasions, if it happened that our talks would run on beyond the appointed time. Then we would go to a huge room where there was a billiard table and a fridge. He would serve me things from the fridge and would produce a bottle of wine, though he never drank a glass himself. Baudouin drank only water.

He was concerned about his heir. Prince Philippe was his chosen dauphin. He had asked me to inform the eldest son of his brother Albert about the workings of the state. Baudouin was firmly convinced that he would become a very old man. As a result Albert would never be his successor. His nephew would soon marry and be sufficiently mature to continue the line without interruption.

Of course he was also aware of Albert and Paola's marital problems, which did not lead to definitive estrangement, thanks to the efforts of my predecessor Prime Minister Gaston Eyskens. I am aware that Baudouin feared indiscretions concerning his brother's illegitimate daughter. I am convinced that, as far as he was concerned, if the news was leaked it would have damaged the integrity of the monarchy. It would also have rendered difficult his brother's succession to the throne.

The development in his ideas regarding succession along the female line was also quite striking. Initially he had let me know that he was against it, but only a year later he in fact came out in favour of it and insisted on it. He was probably afraid that something might happen to Prince Philippe, who was unmarried at the time.

Baudouin played to the fullest his role as King during the last ten years of his reign. His visits and addresses became more striking and also more important. He gradually became a beacon of light for society and stood up for those in need. However, more than forty years of rule had left their mark on Baudouin the man. He was a man with deep personal convictions but one who acted conscientiously with regard to the system within which he worked. He knew exactly what he wanted. Queen Fabiola had a saying about this. When I was waiting with the Queen at Melsbroeck airport, she pointed to the blades of grass growing between the cracks in the concrete. "Just like Baudouin", she said, "soft on the outside".

The fact that Baudouin was deeply religious and that he was driven by his convictions does not mean that he had no respect for other religions or systems of life. He rejected such insinuations out of hand. The greatest moment in Baudouin's life was undoubtedly Pope John Paul II's visit to Belgium in 1985. The Pope's visit was prepared for in the minutest detail. Each member of the royal family was given a role to fulfil in this historic event. Baudouin saw to it personally that everything ran perfectly. After he said farewell to the Pope, following a flawless and indeed enthusiastic reception in our country, the King was so relieved and pleased that he drove around with me for an hour along the avenues and laneways of the royal grounds in Laeken.

My last conversation with Baudouin was in Ghent in March 1993. Queen Fabiola was present along with my family. The King knew at the time that things were not going well between me and my wife. He was much concerned by this. I was seated in one of the front rows during the Te Deum for the National Holiday on 21 July 1993. When Baudouin was leaving the cathedral he looked in my direction. I shall never forget the expression on his face. It was one of concern, one of understanding. In that short glance a period of my life came to an end.

At midnight on 31 July I arrived at the offices of Médecins sans Frontières in Nairobi on my way to Somalia. The next morning I was listening to the overseas news on Flemish public radio when I heard that King Baudouin had passed away the evening before while on holiday in Spain. I felt a lump in my throat and everything around me fell silent. A kindred spirit had passed on.

Life Is a Big Mystery

As a twelve-year-old I had not the slightest notion that the King would play such a prominent role in my life. Neither did I realise how much the war years had influenced my life and would become one of the pillars upon which my commitment to Europe was built.

It was a sunny day, 10 May, when the Second World War began in earnest in Western Europe. On that day I was on my way to kindergarten; a girl was taking me there on her bicycle. I can still see the Langendam, the long road between our neighbourhood and Sleidinge village. Suddenly the girl shouted: "Look up there! Look at all those aeroplanes!" They were German bombers. Germany had invaded Belgium the night before and they were bombing behind the lines.

The invasion was followed by days of confusion. Our little stables were commandeered by a platoon of Belgian soldiers, who needed them for their horses. When the German artillery started firing at them, panic broke out and our mother hid us – four sons, two of whom were twins barely four-months-old – in a pit my father had dug to store the turnips and potatoes in winter. The horses broke loose and vanished beyond the farmyard gate. The Belgian soldiers were not anywhere to be seen. An eerie silence fell over the countryside. Soon after, the Germans drove into our farmyard in their tanks and armoured cars. They stayed there for several weeks. I can remember how one of the German officers berated my mother because my three-year-old brother had tried to open a strange-looking box. It was *schrecklich* because the box was full of hand grenades! Peace returned to our neighbourhood when that first column of German soldiers left. Then began the anxiety and the search for food.

We had enough fruit, butter and milk and even a little bit of meat. Every now and then I would bring flour to the baker a little further up the road and he made better bread from it than the ordinary stuff you would had gotten in the rations. There was a constant flow of visitors to our house: uncles and aunts, nieces and nephews, all from Ghent. They smuggled anything edible they could get hold of. Those images of war, the misery of so many, the total impoverishment of the country, the deep wounds inflicted on our moral fabric through the collaboration and the ensuing heartless repression marked me and the majority of my generation. It was obvious that we would get behind any idea that would unite a Europe that had been torn apart by two world wars. That obviousness seems to be less present in more recent generations. At European summit meetings I meet people who have never experienced war at first hand and, as a result, draw fewer urgent lessons from those dark times.

Each new generation is marked by a series of shared events. Older politicians can bear witness to how important their youth was for their subsequent political action. I also wish to show this through these memoirs.

Chapter II

Conviction and Responsibility

Anyone who examines my career as a politician will probably discover a clear trend running through it: I was a Flemish federalist activist and a student leader, then President of the Christian Democratic youth organisation, Prime Minister of Belgium and after that President of the European People's Party. It seems as if the positions followed each other in easy succession, but in practise these moves were far from obvious. Great resistance had to be overcome with each new step. Chance, personal choices in studies and commitments, power of conviction and changes in the *Zeitgeist* all played an important role in this.

I took my first steps in active politics during the golden sixties. Complex processes of emancipation were taking place all over Europe. Widespread student protests were accelerating social, political and cultural change. Old (power) relations were being broken down, traditional norms and values were being openly questioned. Prominent political figures such as Konrad Adenauer in West Germany and Charles De Gaulle in France disappeared from public view. One had the impression that the end of an era was near.

A similar process was also running its course within Christian Democracy in Belgium. In the latter half of the 1960s a continuity that had been there since the foundation of the party in 1945 came to an end. A change of the guard was at hand, as was the case in other European countries. Old political stalwarts like Prime Ministers Gaston Eyskens and Theo Lefèvre had held a firm grip on the party reins ever since the Second World War. As old men they had reached the end of their careers and, moreover, had fallen behind in the polls. Proof of this was the successive electoral defeats our party had suffered since 1958.

My party not only needed a new generation to replace the old, it also, above all, needed a new image. This had been blighted by issues that had aroused great passion, such as the repression of those who had collaborated with the Nazis during the Second World War; the so-called Royal Question (whether or not King Leopold III should step down); and the conflict about the independence and state funding of denominational, mainly Catholic, schools. The party had been weakened internally through the years because it lacked a clear vision. With respect to the necessary reforms of the country's institutions, it did little more than make lame statements or put forward half-hearted solutions. Slowly but surely, the Christian Democrats were losing touch with the mainstream electorate in Flanders. If the party wished to survive, it desperately needed to come up with a new project. The party leadership slowly began to realise that they would need new blood to take on this task.

A New Generation

In the spring of 1967, I was elected as the sixth national President of the Flemish Christian Democratic youth organisation. I now fully realise that I joined the Flemish Christian Democrats at the right time. The need for commitment from young politicians who could attract voters provided me with the opportunity to break through in the party and to propagate my federalist ideas unhindered. I was part of a new generation of young adults, with a new mindset, who were in favour of radical changes. The word "continuity" was not in my vocabulary. We were the first generation to have been emancipated through education. We lived in more comfortable circumstances than our parents had, and this allowed us to use our creativity in all sorts of ways.

I still clearly remember the first time I met Jean-Luc Dehaene, who would later succeed me as Prime Minister. I was putting together the new executive committee and was looking for energetic people who could take on leadership roles and who were in touch with the youth. His directness of speech and his non-conformity made him the right person for the job. He was soon to gain a reputation within the party for being somewhat of a blunt instrument. He was a non-conformist even in the way he dressed. A far-reaching partnership developed between Dehaene and me immediately after he joined the youth wing of the Flemish Christian Democrats. Both of us contributed considerably to our new project: he wrote numerous tracts and I supported these points of view when addressing others outside the movement. The journey we were to undertake we would share together. For years we formed a close team until political circumstances in that crisis year of 1992 caused us to part ways.

As president of the youth organisation, I set myself the task of carving out an independent space for our movement within the party. In keeping with the spirit of rebellion of the times, I refused to lapse into becoming a well-behaved servant and admirer of the party bosses. One of the first things we did was to challenge the institutional programme of the then unitary Belgian Christian Democratic Party (CVP/PSC) – which did have a Flemish and Walloon section, nonetheless. In the mid-sixties, the party and its doctrine were far removed from federalist thinking. Its programme went no further than the basic principles of decentralisation and a limited amount of cultural autonomy. The principle of a centralised state was in no way questioned.

Promoting Innovative Ideas

On 10 May 1967, we launched our much-debated "Manifesto on Autonomy" (*Autonomiemanifest*). Even today, I am struck by its resemblance to my speech five years earlier to the Flemish People's Movement (Vlaamse Volksbeweging). It is proof for me of the fact that despite joining the Christian Democrats, I have always remained faithful to a deep-rooted Flemishness in me – something I continue to cherish.

The circumstances were also in our favour. The confrontation between Flanders and Wallonia about the division of the Catholic University of Louvain into two autonomous universities, one Dutch-speaking, the other French-speaking, had reached another critical stage. Practically all the Flemish parties were behind the idea that it would be advisable to move the French-speaking part to Wallonia in order to combat the influence of French-speaking students and professors in Louvain. Besides, student numbers had increased to such an extent that a new campus was needed to accommodate them anyway. As a result, Louvain-la-Neuve was founded in Walloon Brabant. Deep-rooted differences of opinion on that matter also caused the CVP/PSC to shatter. Since then the Flemish- and French-speaking Christian Democrats have gone their separate ways as autonomous parties.

It was our desire in our manifesto not merely to unleash passionate debate on reforming the state. We were also in search of concrete political results. For the first time in its history, the Flemish Christian Democrats had a faction that put forward federalism as a solution to the problematic issues between Flanders and Wallonia. Our approach was innovative in that we were moving beyond a purely language-oriented policy and taking steps in the direction of a more structural and institutional solution. Because we formed a highly active minority, we did not remain trapped at the margins and could force the party to focus its interest on federal matters.

The elections of 31 March 1968 ended in defeat for the CVP, just as the 1965 elections had done. Once again, language-based parties like the Flemish nationalist People's Union (Volksunie) and the French extremist Francophone Democratic Front (Front Démocratique des Francophones) made huge gains. The Eyskens government that then took power – this time a coalition of Christian Democrats and Socialists – took up the challenge of seeking a structural solution to the Flemish–Walloon issue.

In that same turbulent year of 1968, the youth organisation of the Flemish Christian Democrats reached a fever pitch in political party terms. We were awash with all sorts of influences and were in the grips of the widespread spirit of rebellion. Many new and interesting experiments were taking off in other countries, such as D'66 in the Netherlands, a party striving for a radical democratisation of society, and also in the Netherlands the Political

Party of Radicals (Politieke Partij Radikalen), which had broken away from the Catholic People's Party (Katholieke Volkspartij). D'66's critique of political vagueness exerted a tremendous attraction for me. They formed a means of escaping from the heavily paternalistic doctrines of our party leaders.

This led to ideas that would see the light of day several months later in our manifesto on the party system in Belgium and on the role of the Flemish Christian Democrats within it. The content of the manifesto, which was issued officially on 11 January 1969, was revolutionary. Our first statement read as follows: "We are of the opinion that our government institutions are in a lamentable state of repair. Not only are the structures in Belgium flawed, party politics is also on the wrong track. It is impossible to pursue innovative politics within the present political divisions".

In order to get beyond the deadlock, we suggested that progressive elements from the various political parties be grouped together to form a "Radical Progressive People's Party", analogous to the Dutch Political Party of Radicals. How could we reach that goal? The first concrete step towards cooperation was that the Christian Democrats would negotiate with the Socialists and reach agreements on electoral policy ahead of the following elections. At the same time, thinking within the CVP had to be radicalised in order to turn it into a party with a genuine programme. In the long term, it would make room for a Radical Progressive People's Party but this development was not to be rushed, and public opinion had to be continuously tested and prepared.

These were daring proposals and were met with clear opposition from the party's central committee, of which I was an elected member. This did not upset me, however. In a speech at the party congress of April 1969, speaking on behalf of the youth organisation, I stressed that "shared religious affiliation was no basis for forming a party and that only a vision for society could meet the need. This vision could only be progressive, as it had to be based on a sincere concern for and commitment to our fellow humans, social justice and peace".

I asked the party for openness towards approaching and working together with progressive Socialists, if that should prove possible or desirable. Much to our surprise, participants at the conference voted in favour of our proposal. The party congress made it a priority that rather than continuing to expand solely on the basis of its prior achievements within Christian Democracy, the CVP should become a programme party whose political action would be based on a progressive vision for society.

On 1 May 1969 – Labour Day – only a few days after the congress, the elderly leader of the Socialists, Leo Collard, issued a striking call to Catholics. He had come to the conclusion that divisions in political opinion along the lines of religious belief were a thing of the past and that all progressive forces should join in forming a front against obsolete politics in Belgium. The welcome Collard's call received from the executive committee of the Christian Democratic youth organisation was clearly positive. It was completely in harmony with what we had already argued for in our second manifesto. My heart pounding, I subsequently paid two visits to Leo Collard to see what common form of action we could take.

However, it soon became clear that the party leadership was far from pleased with the contacts I had made with the Socialist leader. Though I found him to be an honest partner in conversation, our contacts finally came to nothing. In fact, Collard was quite isolated within his own party and was not powerful enough to move them in the direction of forming a common progressive front. To our great disappointment, we realised with time that the majority of the Socialists were merely opportunistic and only regarded the progressive front as a means of absorbing the Christian Democrats into their own party. The initial idea of forming a tandem of two equal political movements that could maintain their own political identity did not seem feasible at the time.

While our ideas about forming a common progressive front were experiencing nothing but resistance, a third manifesto saw the light in July 1969. The document entitled "A Creative Approach to the Reform of the School Pact" (*Creatieve aanpak bij de herziening van het Schoolpact*) caused a stir almost immediately. In advance of the reform of the Pact in 1970, we formulated a proposal to create a new pluralist type of school, one in which the various religious affiliations and philosophies could flourish side by side. As we saw it, the changes in mentality that had taken place in the late sixties and the strong desire for democratisation meant that we were obliged to create a new critically necessary kind of education, one committed to preparing youth for participation in a modern society. The strict divisions in the field of education and among the various education networks had become obsolete and stood in the way of forming a new, open mindset among students. We wanted to resolve the situation by integrating Catholic and State schools with time into a single network of pluralist, mixed, community schools.

Our proposal for reform of the education system brought more hostile reactions, contempt and mocking than even our second manifesto had. People within the Christian Democratic Party were open to internal discussions on federalism and party reform, but tampering with Catholic education and the Catholic block would remain anathema for the time being. The draft document was rejected out of hand and in the sharpest way possible condemned by the whole Catholic establishment, including the leadership of the Catholic school system, the Catholic press and the Christian Labour Movement.

Intellectual Roots

Looking back on my four years as chair of the youth organisation of the Flemish Christian Democrats – the executive committee I presided over became known as the "Committee of Stars" (*Wonderbureau*) due to the large number of members that became senior politicians afterwards – I cherish the intellectual heritage that we left behind in the form of three manifestos. Each one of them drew directly on the three waves my party, my generation and I had ridden in on: Flemish autonomy in a federal Belgium, cooperation with progressive non-Catholics, the dissolution of a polarised system and the founding of community schools. These were generous documents written by a generation who experienced the fact that they too could push for innovation within Christian Democracy.

I am still struck with awe at the fact that I was able to bring together and form a close team with so many talented young adults who had such special qualities and gifts. In contrast to today, during the sixties there were a greater number of people who were prepared to commit themselves politically, even though they were not quite prepared for politics. I had always hoped that after 1971 new "Committees of Stars" would spring up and that each would take the party by storm, establishing ideas and ushering in the winds of new and vital change. The chances of the survival of Christian Democracy have always been intricately bound up with the human potential of the youth. But the pool politicians can draw on has become much smaller nowadays. Those young people who do get involved in politics must realise that the ground on which their thoughts will fall is not as fertile as it was in the past. As a result of this and other factors, present-day youth organisations are prevented from becoming "Committees of Stars".

The central question in all of this is whether a generation of politicians can reach the very centre of political power without betraying their own ideals. In this respect, I was deeply impressed by a Dutch collection of Paul Ricœur's essays, entitled "Politics and Faith" (*Politiek en Geloof*) that was published in 1968. I read and reread them. I would like to quote Ricœur at length here, because he has had a profound influence on my political activity. In the chapter "Requirements for Political Training", he considers the relationship between ethics and politics:

As we know this relationship is difficult and deceptive. Allow me to state at once which working model I [Paul Ricœur] am using, a model I keep to, moreover, as a yardstick in my own personal life.

What it involves is an extremely fertile distinction, borrowed from Max Weber, that great German sociologist from the beginning of the twentieth century. In his famous monograph *Politics as a Vocation* (*Politik als Beruf*), he makes a distinction between two levels of moral behaviour: "the ethics of conviction" – *Gesinnungsethik, morale de la conviction,* as he calls it – and "the ethics of responsibility" – *Verantwortungsethik, morale de la responsabilité*. It is certainly worth noting that in his manuscript Max Weber first wrote of "the ethics of power".

This specification is of great importance for what follows, for it is my conviction that the welfare of the community ultimately rests on a correct relationship between these two forms of ethics. On the one hand, we have an ethics of conviction which is borne by scientific, cultural and religious associations and communities, including the churches, who have their own contribution to make at this level and not in politics.

On the other hand, we have the ethics of responsibility, which is also an ethics of the practise of power, regulated violence and accountable debt. To my mind, it is vital to maintain the tension between these two in learning politics, for if we conflate the ethics of conviction with the ethics of responsibility, we will relapse into *Realpolitik*, into a form of Machiavellianism that stems from a continual confusion of ends and means. On the other hand, if we allow the ethics of conviction to meddle in the other we will only succumb to the numerous illusions of moralism and clericalism.

I have repeated the above sentences hundreds of times in speeches and in my writings. They became my *Leitmotiv*. In using this quote I continued to assert that a Christian Democratic party should never be a party where nothing else is important except power.

Leader of the Flemish Christian Democrats

During the 1968 to 1971 parliamentary term, politicians completely overhauled the institutional make-up of the Belgian state. On 18 February 1970, Prime Minister Gaston Eyskens presented reform of the state to the Parliament in the following stirring words: "The unitary state, including its structures and modes of operation, set out as they are in law, has since become obsolete. The communities and the regions have to find their place in the new structures of the state which are better adapted to the current situation particular to our country". This was translated in the French-speaking media as "La Belgique de papa est morte" (The Belgium of my daddy is dead). Prime Minister Eyskens guided Parliament through the various stages of constitutional reform, the first since the proclamation of the Constitution in 1831. Steps were being taken ever so carefully in the direction of a federal state. The emphasis lay on cultural autonomy, which, in order to maintain ideological balances, excluded education. Other changes rendered regional economic decentralisation possible.

Despite the positive results achieved by the Eyskens government, the Flemish Christian Democrats lost the elections in 1971, just as they had done in 1965 and 1968. This made the need for party renewal even more urgent. The CVP went in search of a new party leader at the beginning of 1972. Four names were put forward and, much to my surprise, one of the names mentioned by a group of fourty-year-olds was mine. The news gave rise to numerous, mainly critical, reactions in political circles. The French-speaking media were openly hostile. Even Manu Ruys, editor of the highly influential Flemish newspaper *De Standaard*, wrote, "Does Wilfried Martens still believe in Christian Democracy as a valid formula that occupies the political middle ground between Socialism and Liberalism? At one time he created the impression in his discussions with [Socialist leader] Leo Collard on the formation of a progressive front that he considered the CVP as a transition to such a progressive grouping. Does he still think so? And if he does, will the majority of the party go down that road with him? And if not, what are his views today?"

Because I was aware that these issues were very much on everybody's minds, I answered him in an interview that "we did wish to form a front but under one important condition, namely that the Socialists had to voice a similar desire for reform. For a long time 'Collard's Labour Day call' gave us the impression that there was something moving in his party. But all that is past tense now. The willingness of the Socialists to negotiate and implement change has completely disappeared since Collard's departure. All his successors have in mind is to humiliate the CVP and to isolate them. It now seems unlikely that any agreement can be reached with the Socialists in the way Collard intended or in the way stated in our manifesto."

Herman Deleeck, a well-known Flemish professor and the ideological mentor of the younger generation of Christian Democrat politicians at the time, later stated that

"The scales had fallen from Martens' eyes and he realised that the daring, theoretically sound positions he had defended along with his team could in no way be achieved and that he could forget about trying to sell them in his own circles. He has foresworn this part of his youth. Others have done so too but far less spectacularly, of course. It is pointless to call his decisions betrayal or calculated. This is the road to adulthood and to taking real responsibilities. The role he has fulfilled since then proves that he knew intuitively which path to take. Martens is not a man of revolution. He will try to push his ideas through, he will even fight for them but he will not needlessly die for them".

"We Can Win Again!"

During the party congress on 4 March 1972, I was elected President of the Flemish Christian Democrats with a total of 83.6% of the votes. I ended my acceptance speech with the following words:

You have elected me as leader, a man whose parents worked with their bare hands. A man now, but one who, by the age of seven, had lost his father in the middle of a war; whose mother carried the load alone through all those years and brought up a family of five until she too died even before her children could find a new home. And yet all five of us have had opportunities in this life that were equal to our talents. All were helped by a teacher or a local priest, who showed us the way to further education, through grants that allowed us to study; through the inspiration of an eminent teacher, a chaplain, a union leader or a politician. This is Christian Democracy alive in Flanders.

Today, we have among us the most experienced Prime Minister since the war. We have the finest team of ministers, new young party activists, an active and effective youth movement, and thousands of convinced grassroots activists who are always ready to come to the aid of the party. Let us close ranks. We can win again!

My job as young president of the party was no sinecure. Even though the unitary party had already been split for four years, since the crisis over the Catholic University of Louvain, the whole infrastructure gave the impression – and many continued to cherish the illusion – that this was only a temporary state of affairs and that the two wings of the party would soon begin to work closely together once again: there was still one president for the whole CVP/PSC, Robert Houben, a national secretary, a joint research centre and a national treasurer. They all had staff and secretaries, many of whom looked back nostalgically on the good old unitary days and also stated so loudly. They considered me somehow as the successor of the earlier heads of the party wings. I hardly had anyone to help me carry out my work.

On the Monday after my election as party president I set up office in the Rue des Deux Eglises (Tweekerkenstraat) in Brussels, and from that day on I committed myself to the party 24/7. I had to hand on my legal practise, which had just begun to be successful, to my young colleague. At that time I already had a blueprint in my head for how to run the party well. I set three priorities: training in politics for representatives and staff; modern, scientifically based communication; and the full participation of women in the running of the party. In order to bring this about, we needed to have our own budget to pursue specific policies. However, party financing depended mainly on complex, impenetrable forms of fundraising and allocation. The impression on the Flemish side was that a disproportionate amount of the funding went to the French-speaking Christian Democrats. I carried out some changes to this situation by appointing our own fundraiser. This certainly was not appreciated, and I even had to face a negative publicity campaign as a result.

Respect for Life

Experience and observation had taught me that the strength of a leader lies in the energy and unanimity of the party he is leading. If I did not want to end up as a lame duck at the negotiating table, then the CVP had to become stronger and Christian Democracy had to develop itself as the leading movement in Flanders.

It did not worry me that we could not reach agreement with our French-speaking party members on Brussels or other language issues. On the contrary, this was one of the reasons why the party had split and why each wing could voice their community's standpoint. In other areas like education, social affairs and ethical issues, however, it was important that we saw eye to eye.

One of these ethical issues was abortion. A few months after my election as party president, I organised an open party day on this morally delicate issue. It was time to reach a decision. And we did, all Christian Democrats together during a common press conference. What is remembered from the press conference is that we found the liberalisation of abortion inadequate. But our position contained a number of measures and proposals that were highly progressive in contrast to mainstream thinking in Catholic circles at the time. We must not forget that condoms were still forbidden by the Catholic clergy and that the Bishop of Ghent, Leonce Van Peteghem, stated after our press conference that "under no circumstances could a child be killed to save the mother".

We proposed that clauses prohibiting the use of contraceptives be removed from the criminal code, and we called on family doctors to inform their patients about their use. The statute on children born out of wedlock was also thoroughly revised. We wanted there to be more help from the government for mothers in need: a considerable increase in baby bonuses and children's allowance for mothers heading single-parent families. On the subject of abortion itself we stated the following:

> The CVP/PSC holds that the problem of abortion cannot be approached without firstly bearing in mind that fundamental unalterable rule of civilisation, that is, respect for human life. Despite modern methods, abortion is an act whose purpose is to cause psychological and physical harm.... Abortion is, therefore, an unacceptable means of regulating birth. Abortion should no longer be punishable when it is carried out under very serious conditions, namely when continued pregnancy would be of serious consequence to a woman's health. In order to guarantee legal security, the law should be amended accordingly.

To conclude we called for "a calm, wide-ranging debate in Parliament in order to arrive at a policy that would lift the community to a level worthy of humanity. This is why the CVP/PSC considers the liberalisation of abortion as inadequate".

A few months later, when the Socialists had submitted a white paper proposing the far-reaching liberalisation of abortion, I stated in my hometown Ghent that the CVP/PSC would utterly reject this proposal:

> When the weaker confronts the stronger, complete freedom leads to oppression; then the law ensures liberation. These principles are so obvious and generally accepted that I am surprised that we Christian Democrats stand alone politically as far as the rights of the child and the mother are concerned. As never before, our society possesses the means to include each human being and to guarantee that all have a dignified human existence.

And now suddenly the willingness to use these means to the fullest is missing. This I cannot accept. A mother in need cannot be abandoned to her fate; no handicapped person can be rejected, the life of each conceived, though yet unborn, child must be respected.

I still stand behind these words. It was an ironic twist of fate, therefore, that in 1990, as Prime Minister, I had to sign an abortion law that I had voted against in Parliament but which had been passed by the majority. The reason for this was that King Baudouin refused to sign the law and so, in order to save the monarchy, the government had to do so. Freedom of conscience and the right to act according to that freedom was not granted to everyone.

Another cause for concern that I wished to raise in the party was care for the environment. Like many politicians of my generation, I was deeply impressed by the report issued by the Club of Rome on the exhaustion of our natural resources and the increase in pollution of the air, water and soil. In my acceptance speech as new party leader I had already warned of the dangers for our children and grandchildren living in a highly polluted world if we were to continue the path we were on. We had to choose either growth resulting in death or balanced survival.

It hurts me to see that such a fundamental aspect of civilisation was ignored for so long, up until the moment that a separate political movement grew out of it, that is, the Greens. The very first open party day that I organised was on the environment. A lot of attention was paid to urban and regional planning. Regional and local authorities needed clear regulations to counter the proliferation of industrial zones and the unregulated expansion in the number of housing estates. This is why regional planning had to be agreed upon as soon as possible. These plans were also advocated in order to stop chaotic forms of urbanisation. Already at that time we drew up plans to control land prices and also pleaded for the transfer of green zones and recreation areas from private to public ownership at a fair price.

It should be noted that these positions were drawn up five years before the arrival of the Greens in 1977. Given their and our concern for the proper management of the earth's resources, the Christian Democrats could also have become an ecological party, without lapsing into a sort of green fundamentalism. Unfortunately, our credibility was sorely undermined by the way these regional plans were bungled, which also involved some policymakers from among the Christian Democrats. It was also the case that because of an economic recession and the slow recovery, our concerns for the ecology were placed on the back burner.

Leo Tindemans

One of the most delicate tasks a party president has to carry out is not only to guide the party to victory – something I have succeeded in doing on various occasions – but also and more importantly to convert election results into political power. It has also become the tradition in Belgium that party presidents, and not the *formateur*, or future Prime Minister, play a prominent role in allocating ministerial posts. Of course, they have to take into account regional and other balances but their opinion is of great importance. It is also to this fact that they owe much of their influence and power.

My first negotiations in forming a government, which led ultimately to a short-lived government of national unity (January 1973–January 1974), were a real ordeal for me. As I would say later on in an interview, "I experienced the old Socialist guard as being exceptionally hateful. The unitary Socialist leaders thought that their moment had arrived, that they could bring the (linguistically) divided Christian Democrats to their knees". We also had another reason for such an exceptional cabinet. The School Pact was to be reviewed. It was very likely that an agreement could be reached only by bringing all the parties who had signed the School Pact into the government.

Leo Tindemans, who had gained considerable power within the various Eyskens governments and who had won the 1971 elections in Antwerp, became the leading figure in the CVP and, therefore, the senior Vice-Premier in the new government. The main argument for this was that Leo Tindemans could lead negotiations on the reform of the School Pact. Our expectations were fulfilled. The only positive result achieved by the government was the reform of the School Pact.

The government experienced a difficult birth and the infant would barely survive the year. The cabinet collapsed and died because of corruption scandals, excessive alcohol consumption and fundamental mistrust among the coalition partners. The country longed for a new leader. As party president, I was convinced that Leo Tindemans would appeal to the electorate, so I launched one of my most successful electoral campaigns ever with the slogan "Things will be different with this man".

For the first time in sixteen years, the party took a giant leap forward and Tindemans was elected with an unprecedented number of first-preference votes in his constituency. The subsequent negotiations were far from easy, however. During the first round of negotiations it became clear that the Socialists were too divided internally to form a stable government. Despite initial failure, the *formateur* was able to hold on to his job and announced that he was willing to form a government supported by members of the Liberal Party and members of the so-called language parties: the Walloon Rally (Rassemblement Wallon), the Francophone Democratic Front (Front Démocratique des Francophones) and the Flemish People's Union (Volksunie).

This was a much bigger step for Tindemans than for me. As a federalist, I had no objections to including parties who had campaigned only in one part of the country. Tindemans, however, carried out his task admirably. We, the party presidents, had almost reached an agreement during late-night negotiations when suddenly, in the early hours of the morning, the *Volksunie* leader rather awkwardly tabled a demand for amnesty for those sentenced for collaboration during the Second World War. The French-speaking negotiators regarded this as sabotage and provocation, and the negotiations grinded to a halt as a result.

On 25 April 1974, Tindemans formed a minority government with the Liberals, which was tolerated by the "language parties", who promised to abstain for a while from voting in Parliament. A few months later, the Tindemans government was extended to include three ministers from the Walloon Rally. A provisional form of regionalisation was quickly adopted and, as a result, three regional ministerial committees and three regional advisory councils could be formed. This was the first tentative yet certainly visible step in the direction of federalism. The first meetings of the regional councils were held with much pomp and ceremony.

This period was undoubtedly one of the party's heydays. Because of our very aggressive campaigns against the Socialists, we gained considerable support in Flanders. The local elections of 10 October 1976, which we contested with the slogan, "Because people are important", were highly successful and proved that the CVP could win again. Our party had enough votes to govern three-quarters of the Flemish boroughs either alone or in a coalition. Unfortunately, the Francophone Democratic Front became the biggest party in almost the whole Brussels area and achieved absolute majorities in five of the capital's boroughs. This French-speaking party also began to gain ground in the Flemish boroughs around Brussels. Their success would seriously undermine any further attempts at state reform.

The first Tindemans cabinet fell due to the volatility of the Walloon Rally, a party that disintegrated while being part of government. When they refused to vote the budget through in March 1977, Tindemans threw them out of government, which meant that he no longer held a majority in Parliament. The situation proved untenable. To prevent his government from becoming a minority, he proposed to the King that Parliament be dissolved. Early elections were called for the second time in three years.

Pioneering Work with Lücker

Readers who do not live in my country might get the impression from what has been said so far that I have been involved mainly in trying to maintain order in Belgium, which is after all a small country. Nonetheless, the European dimension has never been absent in my public and political engagement and activities. As President of the CVP, I was given the opportunity in the early 1970s to make my interest in the European integration process more concrete. I was involved in the European People's Party from its very foundation. Yet I could never have imagined just how much the EPP would dominate and give direction to my political life after my being Prime Minister.

The founding of the EPP was not something that came out of the blue. There had already been a long tradition of international cooperation within Christian Democracy in Europe dating back to the 1920s. A prominent role was played by Flemings, all of whom had an international focus. From 1950 till 1959, the founder and President of the CVP, August-Edmond de Schryver, led the Nouvelles Équipes Internationales (New International Teams), as the group of West European Christian Democratic parties were known at the time. Our party president and later Prime Minister, Theo Lefèvre, led the Equipes from 1960 till 1965. The Équipes were succeeded by the European Union of Christian Democrats (EUCD), which was founded in 1965, with Leo Tindemans its first Secretary-General. He held the position until he became Prime Minister in 1974.

A Christian Democratic group had already been set up within the European Parliament in 1953. Our group wished to found a European political party in view of the first direct elections to be held in 1979. The first real steps in that direction were taken within the Political Committee of the Christian Democratic Parties of the Member States of the European Community (EC). As newly elected President of the CVP, I was a member of the Committee along with my colleague from the PSC, Charles-Ferdinand Nothomb. I never missed a single meeting. A total of eleven parties were represented, from seven Member States of the European Community: the CVP/PSC from Belgium; the Dutch ARP, CHU and KVP, who would later merge to form the Christen Democratisch Appel (CDA) in 1980; the CSV from Luxembourg, represented by their new party president, Jacques Santer; the CDU and CSU from Germany; the Italian Democrazia Cristiana; the French Centre des Démocrates Sociaux, which became part of the UDF from 1978 on; and Fine Gael from Ireland. As they did not have any Christian Democratic parties, Great Britain and Denmark were not represented.

Hans-August Lücker, a member of the CSU from Bavaria and the then head of the EPP Group in the European Parliament and also Vice-President of the EUCD, was very much in favour of long-term planning for Christian Democracy in Europe. To do so, a stable political formation was needed, that is, a party. To work alongside Lücker, a man more than twenty years my senior, who had considerable experience in European affairs, I was appointed *rapporteur* for the ad hoc working group on a "European party" in 1975. The working group met on several occasions between November 1975 and January 1976 to discuss the constitution of a future European party. A draft of the statutes was presented to the Committee during a meeting in Paris, which it then passed during an official meeting on 29 April 1976. This constituted the actual founding of the European People's Party – Federation of Christian Democratic Parties in the European Community, as our European party was known in full back then. The official foundation took place later, on 8 July 1976, in Luxembourg. Leo Tindemans, who had since become Prime Minister of Belgium, was elected the first President of the party.

Setting up the EPP proved to be a very intense and arduous process. We had to hurry because the Socialists, to whom we had lost the relative majority in the European Parliament since Great Britain's membership, had already founded their Federation of Socialist Parties in the European Community in 1974. The Liberals had also wasted little time in setting up the European Liberal and Democratic Federation in 1976. However, there was considerable discord within our ranks about the scope of a prospective European party. Would it be open to other non-Christian Democratic parties? It was mainly the Germans who insisted on this, as they had maintained good relations with the British Conservatives in the past and thought it unimaginable that such a party would have no representation in one of the largest Member States of the European Community. According to the CDU/CSU, only a permanent coalition of Christian Democrats, Conservatives and Liberals, in keeping with the German model, could offer any form of opposition to Socialist dominance in Western Europe. In the post-May 1968 period Western Europe was predominantly red in colour. Socialist parties were in power in West Germany, Great Britain and the Netherlands, among other countries. The CVP/PSC was the only Christian Democratic party capable of winning parliamentary elections in Europe in the 1970s.

Together with the French, Dutch, Luxembourg and Italian Christian Democrats, we were against forming a front with the (British) Conservatives. Quite to the contrary, many of us were in favour of working with the Socialists in national politics. Because I personally belonged to the latter group and because my colleague Lücker represented the German stance, we formed in ideal team for bringing about a compromise. We both attached great importance to the development of an integrated Europe, the ultimate goal that had always to be kept in view. To a certain extent, the founding of a European party was only a means to that end. But that meant it had to be based on sound ideological foundations and also be sustained by a long-term vision. If Christian Democrats wanted to remain of significance in Europe, then cooperation with like-minded parties could not be excluded, at least not in principle.

Vision and Initiative

This view was strongly criticised by many, not least many within my own party, particularly when I as EPP President led the negotiations to expand the EPP and include other non-Christian Democratic parties. I was reproached then and am still regularly chided for having made an ideological U-turn. It would appear that at the beginning of my political career, I was in favour of forming a progressive front, whereas towards the end of my career, I have devoted myself within the EPP to developing the "European right". I wish to deny this in the strongest possible terms.

I have always remained true to the basic principles of Christian Democracy: personalism, a social market economy, subsidiarity and European federalism. I have anchored these principles within the EPP and it is only because like-minded people from other parties have subscribed to these principles that they were allowed to join the EPP. Opting to be a small, select, but powerless club of "pure" Christian Democrats, or to be absorbed into a larger formation in which Christian Democrats would form only a minority and hardly be capable of realising their ideas, have never been options I have considered. It is from this point of view and conviction that my commitment to the EPP should be understood. We need a strong, broad-based EPP to continue to strive for a united Europe, next to and along with the Socialists and the Liberals. Our real opponents in the European project are the extreme right and the increasing number of populists and Eurosceptics.

The awareness that "pure" Christian Democrats would lack punch at the European level and the conviction that the strength of Christian Democracy lies in its platform and that its ideas can form the basis for political innovation have always formed the basis for my political action. I have always been filled by this vision, both now as EPP President and at the birth of the EPP, as the following note I wrote in 1975 about the foundation of a European party bears witness:

The formation of a popular front between the Socialists and Communists on the one hand and the far-too-sharply posited polarisation between Socialists and Christian Democrats on the other hand could lead to a political division of Europe which would seriously jeopardise the process of integration, or even render it impossible. Cooperation between all anti-socialist forces is fraught with risk because it is insufficiently founded both ideologically and programmatically. This could be seriously detrimental for the future development of Christian Democracy, which itself relies on political principles, and also for a European party that will need its own ideological principles, if it is to maintain its identity and safeguard its future. The Christian Democratic parties within the European Community will have to take the political initiative to ensure that the political union of Europe rests upon the broadest possible foundations, that Christian Democrats work together in a European political party and that this party keeps open the possibility of working with all democratic parties.

A European People's Party

The combined effect of a deepening and a broadening, the acceptance of Christian Democratic principles by like-minded political parties which at their foundation had no link with Christian Democracy: all of this lay at the basis of the compromise Lücker and I wished to reach in setting up the EPP. Discussions crystallised around the name and terms of membership of the future party. The matter of the name – whether or not the name should refer to Christian Democracy – was highly sensitive. A plea was made to opt for the name "Democratic Centre". We could not reach agreement and various suggestions made the rounds: European People's Party, European Christian Democracy, Christian Democratic European People's Party and the European Social Party for Progress.

The matter of the name was not merely incidental. A name always indicates which identity you wish to portray. Those in favour of including the Conservatives would rather avoid the term "Christian Democratic". On the other hand, those opposed to their joining regarded this term as a guarantee for the maintenance of the Christian Democratic character of the Party. Only at the last minute was consensus achieved regarding the name "European People's Party", which included a reference to Christian Democracy in the subtitle.

The term "People's Party" met the German demand for openness regarding political orientation and socio-economic class and also referred to the various Christian Democratic people's parties in other countries, like the CVP in Flanders, the KVP in the Netherlands, the CSV in Luxembourg, among others, as well as to the predecessors of the CDS in France and Democrazia Cristiana in Italy.

I have never seen a contradiction between Christian Democracy and a people's party. To the contrary, a Christian Democratic party is a people's party *par excellence*. In retrospect the choice of the name "European People's Party" was visionary. Without this name we could never have broadened the Party.

Unlike the decision regarding the name, the matter of membership was agreed on contrary to the wishes of the Germans. Only Christian Democratic parties from the member states of the European Community at that time could become members of the EPP. As far as the future of the Party was concerned, this was a highly important *Auseinandersetzung*, because it was mainly the Germans and particularly the President of the EUCD, Kai-Uwe von Hassel, who were pushing for an opening up of the Party.

Their position was not followed, however. In fact, parties of other political orientations, like the British and Danish Conservatives, were excluded as well as Christian Democratic parties from non-EC countries like Switzerland and Austria. Because the CDU/CSU nonetheless wished to maintain structural contact with non–Christian Democratic parties both within and outside the European Community, a few months later they set up the European Democratic Union (EDU) in reaction to the founding of the EPP. The EDU was an association of Christian Democratic, Conservative and other non-collectivist parties. Helmut Kohl, the CDU President at the time, Margaret Thatcher, the leader of the British Conservatives, both of whom were then opposition leaders, and Jacques Chirac, the leader of the neo-Gaullists, were the most important figures within the EDU.

In fact, the Germans were taking a huge risk. During the delicate negotiations about the EPP, they had also been preparing to set up the EDU. It is true that the EDU was not a federation of parties and did not form a group within the European Parliament, but its foundation put a damper on the enthusiasm with which the EPP got off the ground. The date of the founding, only one month after the first EPP Congress, was taken as a provocation. During the meeting of the Political Bureau following the congress that had established the EPP, the debate was devoted entirely to the setting up of the EDU. They had created a negative impression of the new EPP and tried to cause internal division. One has no idea today just how strong the divisions were at the time.

The first years of the EPP were years of deadlock. Three associations of Christian Democratic parties in Western Europe – EUCD, EPP and EDU – worked separately and to some degree even against each other. There was no sign or mention of structural cooperation. Mistrust reigned. The inheritance left by Lücker and myself seemed to have dwindled. The means to our goal, that is, the creation of a federal Europe, suffered from a loss of strength at the outset. Moreover, the first direct elections to the European Parliament were won by the Socialists – in terms of seats at least, for the EPP did get more votes. Later the Socialists' advantage would increase also in number of votes.

As Prime Minister I had to let go of my brainchild while it was still under this dark cloud. It was only later, under my EPP presidency, that efforts were made once again to bring about structural cooperation between both Christian Democratic and other like-minded political parties and, to top it all off, the union of the EPP, the EUCD and the EDU to form one organisation.

Christian Democratic Principles

Next to the statutes that to this day form the basis for the running of the EPP, another important ray of light is the common political programme that was drawn up at the founding of the EPP. Political programmes have played an increasingly important role within the Party. They are the outcome of the common ground shared by the parties, which even since 1976 have continued to differ greatly. It is vital to recognise these differences while at the same time looking for points in common. And that commonality lies in Europe.

Because much stress had been laid from the outset on the programme, a special committee was set up. I was the *rapporteur* along with Lücker. In editing our texts we could rely on the manifestos issued by the EUCD and the Christian Democratic World Union. I also brought in my own party's research centre, which coordinated the contributions of the various research units of the member parties. The 1979 election campaign for the European Parliament was headed by the slogan "Together to a Europe of

Free Peoples". In fact we were setting out the boundaries of the election platform. The Christian Democratic roots were clearly visible from the beginning and would be even more in evidence at the foundation of a European federation:

We, the European People's Party–Federation of Christian Democratic Parties of the European Community, desire a united Europe. With this goal in mind we will continue the successful policies of Christian Democratic statesmen Robert Schuman, Alcide de Gasperi and Konrad Adenauer, who laid the foundations for all the successes realised till now. In following them, we are determined to continue their work and to bring it to completion by founding a European Union, which in political terms will attain its completion in a European Federation as set out by Robert Schuman on 9 May 1950.

This conviction regarding federalism was crucial and an extension of what I had been arguing for in Belgium. It was also what I later discovered in Jean Monnet's writings. I consider Monnet to be the inspiration for my belief in Europe. His biography, which I have read and studied in depth, has also been an important source of inspiration for my political action.

At the very first EPP Congress I laid my cards on the table in outlining my vision for the future of Europe:

For us the unification of Europe should result in a European Federation because a federal structure is the only structure that can bring about and guarantee unity in diversity and diversity in unity. For us a federal structure is best adapted to give form to the principle of subsidiarity: only what we can deal with within the larger entity should be transferred to that entity. In this way the federalist structure fits our pluralist view, one that abhors monopolies. In brief, federalism is our "personalism" in political form.

Belgium or Europe?

Our political programme was agreed to at the first EPP Congress on 6 and 7 March 1978 in Brussels. With it my work had come to an end for a while. As Prime Minister, I could not follow the day-to-day business and developments within the EPP, except at summit meetings with EPP government leaders in preparation for the European Council. It was not I but Tindemans who became the President of the EPP, at my suggestion. He was elected because of his premiership and his renown in and services to Europe. It was not for nothing that he was called "Mister Europe". The fact that he came from a small country and lived close to Brussels also played a role.

However, at one time, before I knew that I would become Prime Minister, the hands were dealt differently. Lücker asked me to become President of the EPP because all the member parties he had consulted had put my name forward.

Paradoxically, it was only Charles-Ferdinand Nothomb, the other Belgian party president in the running, who was against it. I also wonder whether becoming President of the EPP was the right step to take. Certainly, Belgian politics would have looked different if I had done so. Ultimately, I advised Lücker to put the proposal to Tindemans. Tindemans accepted and became President. Nonetheless, I remained heavily involved in Europe in terms of policy. Moreover, by that time my interests and preferences seemed to be focused for good on Europe. When a Flemish weekly asked me a few months after the EPP Congress whether I would not prefer to be Prime Minister of Belgium rather than dedicate my political career to Europe, my reply was this: "If somebody does not take a risk for Europe, then it will never come to anything. Yes, I do believe that the European dimension reverberates most strongly in me". I told my friends in the CVP as well that "Tindemans would inspire the party in Belgium and I would mainly concentrate on Europe".

Chapter III

In Charge of a Restless State

The election results of 17 April 1977 were strikingly favourable for the Christian Democrats. The most significant eye-opener was the number of first-preference votes won by Prime Minister Tindemans. He wished to bring together a broad majority to push through constitutional reform. He tried from the outset, therefore, to form a three-party coalition government with the Christian Democrats, the Socialists and the Liberals. The *formateur*, the person appointed by the king to negotiate a new government, was faced with no easy task, however. The Socialists were quick to announce that they would refuse to form a government with the Liberals.

Instead of the Liberals, the so-called language parties – the Flemish Volksunie and the Brussels Front des Francophones – were invited to take part in the negotiations. We took the risk of opening up discussions with these new partners. The Egmont Palace was closed to the media and we all sat around the table together, all of us fired up by the desire to hammer out a definitive solution in the form of a community pact.

An agreement was reached on one of those balmy days of May 1977. But the resulting "Egmont Pact", called after the palace where the pact was eventually signed, was far from being a paragon of clarity. Differences of opinion soon arose in the government. The interpretations of the pact's wording differed so much that the negotiators had to meet again at another castle, called Stuyvenberg, from 24 September 1977 until 17 January 1978, in order to fine-tune certain key sections of the pact. However, that did not put an end to the difficulties. The Council of State, the highest authority on administrative law in Belgium, which acts as an advisor to the legislature, found that essential sections of the Egmont Pact were at variance with the Constitution. So the pact could not be ratified without amendments. All hell broke loose in the media, and tension between the coalition partners came to a head. The pact reached its demise on Friday 4 August 1978, a fatal day that will always remain etched in the minds of all those who participated. A deep, irreparable chasm had opened up between French-speaking party leaders and Prime Minister Tindemans. They had lost all confidence in him; this became evident during an emotional meeting at his residence. The President of the Walloon Socialists, André Cools, accused Tindemans of being *une crapule*, a crook, and of plotting against those who had committed themselves to the pact. Some horrible scenes took place, which participants at the meeting will never forget. Ultimately, we all parted in a state of shock. And so, the Egmont Pact landed forever in the waste-paper basket of history.

My memories of that painful day, 4 August 1978, were probably coloured by another disaster that happened on the same day. The atmosphere created by the confrontation between Cools and Tindemans had hardly eased when I received the news that Chris, my thirteen-year-old son, had been involved in a serious accident on the south Spanish coast. Would this family tragedy now prove too much to bare? I just managed to hold on, but that period was horrific. When it all became too much for me I would go out for a walk on my own and pray to God to save my child. It was the first time in many years that this had happened to me. Many people wondered afterwards why I had to spend that summer in Parliament. I have asked myself the same question a thousand times. Why did I let my wife and children go off on holiday alone? Perhaps it might not have happened if I had been there. I have never stopped feeling guilty about the tragedy.

Following the Egmont disaster, the Christian Democrats emerged as the strongest party in the elections of 17 December 1978. We had to provide a *formateur*. Tindemans wanted us to stand united behind him, but none of the other parties were now willing to participate in a government with him. The Flemish Christian Democrats were faced with a choice: stand behind Tindemans to the very end, or sacrifice him for an alternative. Flemish Socialist Willy Claes was appointed as *informateur* - the person tasked with conducting informal talks in order to designate a *formateur* - to explore the terrain. He first let the politicians catch their breaths and come to terms with the election results. He held more than thirty rounds of talks during the final days of that year. On 8 January 1979 he completed the rounds. When I met him on that same day, I asked him, "You are surely not going to put me forward as *formateur* when you speak to the King?" Willy's answer was the one I most feared: "Of course I am, because there is no other alternative".

Completely against my will, I was appointed by the King to form a government. But it was highly improbable that I would succeed. I had the support of the parties who had previously held the majority, but my own Christian Democratic activists were vehemently opposed to my succeeding Tindemans as Prime Minister. The attempt to form a government dragged on endlessly, the crisis lasting almost one hundred days. The King appointed a new mediator, the French-speaking former Prime Minister, Paul Vanden Boeynants, who finally was able to broker the agreement. This agreement also had to be approved by my party and, following a tense congress, I was finally appointed Prime Minister.

16, rue de la Loi

The Prime Minister's office, commonly known as "Wetstraat 16" or "16, rue de la Loi", was not unfamiliar territory to me when I became Prime Minister on 3 April 1979. I was only a youngster of forty two then but I knew the building, as I had been an adviser to the cabinet fifteen years earlier. Situated on the corner of Wetstraat and Hertogstraat facing the Warande Park in Brussels, this valuable historical building is part of a splendid well-preserved neighbourhood built in classical style. Inside its white walls the place literally hummed with political activity.

The main entrance to "N° 16" is a very familiar site to Belgians. It is somewhat like 10 Downing Street to Britons. When I called my ministers to a meeting, chaos would sometimes ensue at the front door. It was not unusual for passers-by to bump into ministers who were dropped at the door by their chauffeurs and who were trying to force their way through the throng of journalists, laden down as they were with stacks of heavy portfolios. Once inside, they then had to wait for the annoyingly slow lift in the ministry. The lift would get stuck regularly and someone would have to be sent up the stairs to free the ministers from their plight. While this was going on, the other ministers waiting below were descended upon by the hoard of journalists trying to tease out some stray piece of sensitive information.

That closeness between government leaders and their voters was typical for our country. After every cabinet meeting, I used to go to the small room on the first floor to inform the press of the decisions taken that day. The room was so small that the journalists could literally look over my shoulder and read my notes. When, in 1984, the ministers became the target of attacks by the terrorist organisation Cellulles Communistes Combattantes (CCC), the whole situation became untenable. Cabinet meetings were moved to other locations outside Wetstraat 16, as the ministers were forced to retire to castles or palaces in the Brussels region like Egmont, Stuyvenberg, or Val-Duchesse - the castle where the Treaty of Rome had been negotiated.

A Referee and a Seeker of Consensus

Christian Democrats held power in Wetstraat 16 for thirty years from 1968 to 1999, with a short hiatus in 1973. The post of Prime Minister is usually held by the strongest political party in the Parliament. As a result, the Prime Minister is almost always Flemish. Between 1958 and 1999, Christian Democrats occupied Wetstraat 16 for 458 months, but no one held the office very long. I myself, by contrast, held the office of Prime Minister for 146 months, almost a third of the period that the Christian Democrats were in power. Following the 1999 elections, Flemish

Liberals formed the majority in Parliament and their leader, Guy Verhofstadt, held the premiership from 1999 to 2008. The Christian Democrats then landed in the opposition. Thanks to the victory of our new leader, Yves Leterme, in the elections of 10 June 2007, he could assume government leadership following painful negotiations in the early months of 2008. It was the longest political crisis our country has ever known. But in December 2008 another political crisis broke out and Yves Leterme had to resign as Prime Minister.

King Albert II asked me to mediate the crisis. As mediator, from 22 to 28 December, I was able to restore confidence between the coalition partners and renew government stability. A new Prime Minister from my party, Herman Van Rompuy, was appointed.

The Office of Prime Minister in Belgium is laid out in Article 99 of the Constitution. The article has a threefold effect. It recognises the cabinet as a permanent body, sets out the terms of its composition and renders constitutional the function of Prime Minister. Since 1970, an equal number of French-speaking and Dutch-speaking ministers have held office in the federal government. The rule of parity excludes the Prime Minister should there be an unequal number of ministers. He or she is considered linguistically neutral, as it were. Secretaries of State, as deputy ministers are called in Belgium, are not subject to the parity rule. In 1993, the number of ministers was limited by the Constitution to fifteen. During my fifth period in office (Martens V), more than ten years before this constitutional limitation came into force, there were also only fifteen ministers and ten Secretaries of State.

Whoever moves into Wetstraat 16 is given a post that brings considerable power with it. This individual controls the political agenda. All initiatives and all their trains, as it were, pass through the Prime Minister's office. The Prime Minister can sort them, let them leave early or leave them waiting at the platform or even shunt them into a siding.

But what does the power of the Prime Minister rest upon? He or she is usually described as the *primus inter pares*. The Prime Minister does not act as a representative of his or her own party. The role is that of referee and seeker of consensus. The Prime Minister is assisted by the Vice-Premiers. They are the most influential ministers chosen from the coalition partners. As a result, it is more accurate to describe the Prime Minister as the leader of the majority. Together with the Vice-Premiers and a few other important ministers, a core Cabinet is formed.

This core Cabinet meets whenever it becomes impossible to maintain the centre of political power within the Council of Ministers. It is often unavoidable that vital political decisions are taken by a small committee. That was not so during my fifth government, Martens V. The level of cooperation was such that my fifteen ministers had no need of a core Cabinet, as it were. This was not the case during my other terms in office. The Friday meetings of the Council of Ministers were prepared in great technical detail, but if a real political problem arose it could never be resolved there. Things repeatedly reached a stalemate, and the Council invariably ended with the conclusion that no agreement could be reached. At that time, the centre of political power lay far too often in the head offices of the political parties. There were far too many experts and powerful persons beyond the walls of Wetstraat 16.

It became important, therefore, to bring the most important leaders of the coalition parties into the government itself. In this way, the centre of gravity shifted to the Council of Ministers. Political problems that arose during Martens V were settled, therefore, in the Council. The decision in 1982 to devalue the Belgian franc was taken by the Council of Ministers, as was the decision in 1985 to station cruise missiles on Belgium soil.

Deciding by Consensus

As head of the Council of Ministers, it is advantageous for a Prime Minister to have a legal mind in order to anticipate any conflict and reduce it to its most important factors. The Prime Minister requires considerable insight into human nature in order to be able to recognise the personalities, psychological makeup and the sensitivities of the most important ministers. One pays dearly for any kind of awkwardness in these areas. The Prime Minister also has various tools at his or her disposal to quell possible conflicts or at least to diffuse them. So it is important to master the art of political agendas and hence create a balanced council of ministers. The Prime Minster must be vigilant and make sure that problems do not pile up. Much of what is discussed in the Council has often already been discussed by the administration. These form the more "innocent" points on the agenda. Next to these, an agenda should contain only two or a maximum of three sensitive issues. If the issue is of vital importance, then it is advisable not to allow any other possible point of contention onto the agenda.

A Council of Ministers is run differently from cabinets in most other countries. In Belgium there is a deeply rooted, almost constitutional custom that a government should reach its decisions through consensus. This rule is usually set aside elsewhere. In the Netherlands, for example, decisions are usually taken by majority vote. This is always the case in Israel. In France, where ministers are not even allowed to bring their own documents to the *Conseil*, conclusions are prepared beforehand and formulated by the President. In Germany, the Chancellor does not need the agreement of his or her council of ministers to take government decisions. If ministers have fundamental objections to a particular decision, they may even be obliged to resign.

The sacred rule of consensus holds sway in the Belgian government. Given that the Constitution sets out the parity of languages in the Council of Ministers, a resignation can automatically lead to the fall of an entire government. This of course further burdens the Prime Minister's task. His or her task is to listen and, if serious differences of opinion arise, to prepare a denouement in the debate. There are various ways in which this can happen. Sometimes I would take note of the points of contention during the discussion and allow a final proposal to take form in my mind, which I would then put forward. The minister could then only insist on where to put the commas and full stops.

To guide the debate, the Prime Minister can also take fewer steps and make a proposal that allows room for amendment and change. And if it proves necessary to reach a final consensus, the Prime Minister can force one or two ministers to accept the proposal or else insist that they step down. In the case of heightened tension in the Council of Ministers, it is sometimes opportune to adjourn the meeting on exceptional occasions. This should not happen too often, however. Adjournment is public proof of a lack of unity in the government and of the fact that it has become a bad team. I was highly adverse to such adjournments. When ministers asked for a meeting to be adjourned, I tried to resist the move. This was also my stance when there was a call to carry on the debate further into the night.

Unionist Federalism

I have always considered federalism as the most favourable working structure for the modern Belgian state, the political method that would allow our country to function with true unity. The resolutions I put forward as a Flemish activist back in 1962 already contained the maxim to which I have remained true all my life: autonomy where possible, strong central power where required. This encapsulates the correct meaning of the basic concept of *unionist federalism*. For the last forty years, it has run like a golden thread through all of my political action. The great task remaining was to create a mindset in which federal loyalty, or what the Germans call *Bundestreue*, would become obvious.

I will now try to draw a picture of the young man I was forty five years ago. I was no agitator, but a restless jurist who delved deeper and deeper into constitutional law on the one hand, and helped organise mass meetings of the Flemish electorate on the other. One of the greatest obstacles facing federalists at the time was that there was no clearly shared view on what a federalist state should look like. For some people in the establishment at the time, federalism was a cover for separatism, whereas others mistook federalism for regionalism or decentralisation. Federalism was even approached in a very rudimentary way by federalists themselves; that is, as a "garbage bin" in which the Flemings and the Walloons could dump their frustrations about the unitary state.

I realised that federalist thinking would gain credence in political circles only if it could rid itself of some perniciously confusing concepts. Beyond this, urgent work had to be done to translate federalist doctrine into realistic political proposals for the state. As I understood it, one essential requirement was that federalism would take place in a Belgian context with a strong central government and with an equally strong sense of national solidarity. My loyalty to the concept of unionist federalism sets me apart from the nationalists. I continue to believe that certain common reforms which have been brought about, like social security for example, should not be allowed to fall by the wayside. And I continue to believe this to this very day. This explains my aversion to renegade forms of nationalism like those currently voiced by the extreme right.

Towards Institutional Adventurism?

On 8-9 August 1980, during one of my periods in office, two important acts were passed which have come to be known as the 1980 State Reform. The fundamental point at the time was that next to the national Parliament and government, a Flemish Parliament and government would be also be founded, the same institutions being founded on the French-speaking side. This formed a real step in the direction of a truly federalist Belgium: communities and regions as constituent states were given extensive powers and high levels of autonomy, as they also had the power to ratify decrees. The powers of the regions were increased considerably to include transport, public works and infrastructure. In 1988-1989, education was also transferred to the communities. The importance of these reforms became visible in the enormous increase in the financial powers of the federal states: they went from being responsible for 10% of the national budget to almost 40% of the total. My successor, Prime Minister Jean-Luc Dehaene, "placed the roof on the house" of federal Belgium. Article 1 of the Constitution of 17 February 1994 sets out that "Belgium is a federal state comprising communities and regions".

The ink of the new federal constitution was hardly dry on the page when it was written off in certain circles as obsolete. As former Prime Minister and state reformer, I followed these developments with increasing surprise. And it was not only the radicalism in the thinking of some that was cause for concern. Others wished to further undermine the federal system but have failed so far to draw up a workable coherent alternative. Despite the current political climate, and much to the annoyance of many, I have refused to distance myself from unionist federalism. I have always taken it as my point of departure that high levels of autonomy have to be granted to Flanders and Wallonia, along with the acceptance of guarantees of centralised policy- and decision-making, if we wish to create a new form of balance in Belgium. An obvious logical consequence of my many years of action for the Flemish cause was that Flanders would be declared a public legal entity with its own policies, including recognition in the Constitution of the integrity of its language and its borders. But I have never doubted for a moment that this autonomy would be realised within Belgium and in the broader context of the European Union. That meant that the autonomy of Flanders would never be total, nor would it become an independent state, but rather a federal state in Belgium and in Europe, that is, a federal state that would have and exercise its own powers also within the European Union.

Such an integration which safeguards the identity of the various federal states within a federal Europe, and which would not be centralised but rather decentralised in keeping with the principle of subsidiarity, corresponds to the course of history. Forming mini-states – *Kleinstaaterei* – is contrary to the historical developments that consolidated the formation of the European Union since the Second World War and the fall of the Berlin Wall. If Belgium can function efficiently as a federal state with three languages and cultures, situated at the centre of the European Union, it will become a huge magnet and undoubtedly also become the heart of Europe. I can sum all this up with the following motto: "Flanders is my home, Belgium my fatherland and Europe my future".

A Global Concept

My federalist convictions are founded on an all-encompassing notion. The notion of unionist federalism forms a whole that is supported by its integral parts, none of which can be ignored for fear of endangering the whole. Its essential elements are these: the genuine autonomy of its federal states, a cohesive federal government, a division of powers and finances between the federal government and the states, the principle that there can be no political power without equal financial responsibility, constitutional recognition of the borders and integrity of the linguistic regions, and solidarity among the constituent members of the federation. These essential elements cannot be called into question as a result of shifts in the political *Zeitgeist*.

The convictions outlined above explain why I have always been and will continue to be an opponent of confederalism. This concept is founded exclusively on the willingness of independent sovereign states to cooperate in certain areas that they themselves stipulate. A unifying vision is nowhere to be found in confederalism, nor is there any real form of cohesion based on common rules. Everything depends on the good will and possible interests of the confederate partners. And one could rightly ask whether there would be any good will left if separation actually did take place in Belgium.

The Civil Rights of a Federal System

The concept of federalism is based on one of the great achievements of Thomist philosophy, the principle of subsidiarity. This can and in fact must be approached in two steps. In the first step, levels of government that are above others are required to serve those below them. One must give preference to the basic community, for it is closest to the people and their concerns. This emphasis on the primacy of the basic levels of government was very strongly present in the minds of the founders of the United States, all of whom were permeated by Christian philosophy and care for their fellow human beings. It can also be found explicitly in the 1931 *Quadragesimo Anno* encyclical: *subsidiarii officii principium*.

In the second step, the point is one of choosing which level is best placed to tackle a given problem. And this is not necessarily the lower level of government. In this respect, the higher authority only intervenes in a subsidiary way on the grounds of complete equality. This means that each level of government is only allocated powers to solve matters it is best placed to solve, given the size, nature and importance of these matters.

Taken together, these principles of subsidiarity and solidarity form one of the basic pillars of European integration. I am also convinced that solidarity is one of the basic pillars of the federal state. The complete devolution of social security to the regions in Belgium, for example, would seriously undermine a vital element of our federal system.

How do I feel about the federal restructuring of my country after forty five years of political commitment? In *The Man without Qualities,* Robert Musil wrote of an imaginary country called Kakanien, "in which, stupidly, there was an aversion of each person to the efforts of every other person, something we all agree on now, and something which at an early stage had been perfected into a sublimated ceremonial way of being". According to Musil, this could have had even greater consequences for Kakanien were it not prematurely halted by a catastrophe. I hope that the Flemings and their French-speaking counterparts will not need a catastrophe in order to learn to appreciate the civil rights of a federal system.

My Heart Is Almost Torn, Writing These Words

The West was hit by an economic crisis in the 1970s. Politicians did not seem to be prepared for the shock of the oil crisis in 1973. The direct reason for the rise in oil prices was retribution for the Yom Kippur War, which the Arab countries had lost to Israel. The shockwave considerably disturbed the economic balance of all European countries. In Belgium, however, the crisis was even more profound and lasted much longer. In the blink of eye, we became burdened with the highest national debt in Europe. And all watched our excesses with amazement. My country became known as "the sick man of Europe", not only among politicians but also within international organisations. They wondered why we were unable to push through a programme of economic recovery. The cause for this lay in the nature of our system, as a result of which Belgium was much slower than the other EU Member States to implement any form of crisis management.

The first cause stemmed from the mindset of the political classes. They continued to fool themselves into thinking that the crisis was only short term and that prosperity was just around the corner. Curbing spending would be unnecessary because it was the government's job to stimulate growth. This poor understanding of Keynesian theory caused politicians to believe that they were behaving well by handing out lots of money. The accumulation of debt was considered an expression of sound, far-sighted economic policy. Moreover, differences between the Flemings and their French-speaking counterparts prevented politicians from devoting their full attention to solving economic problems. As a result, Belgium only came to this rude awakening five years later than the other industrialised countries.

Successive governments tried unceasingly to maintain social peace. They dared not tackle the excesses of collective bargaining. The unions were important players and all social partners were present at the table with the Council of Ministers, ready to negotiate the policy to be followed. When we were obliged to tighten our belts following the eruption of the crisis, tensions rose between the social partners and the government. It was especially the unions who failed to recognise their role in this changed situation. They continued to demand pay raises, even though the recession had rendered this impossible.

Tensions were mainly focused on pegging wages and benefits to increases in the cost of living. A heroic war for linking incomes to inflation, called the index, was waged for years between the unions and left-wing parties on the one hand and the employers and centre-right parties on the other. Because its purpose is to maintain spending power, the index is in principle a just mechanism for the distribution of prosperity. However, it can become perverse if too much indexing is carried out too quickly. This is what happened in Belgium in the early seventies. It caused the crisis in our country to get out of hand.

Leo Tindemans' first government (1974–1977) tried to set economic reform in motion. A hesitant attempt to limit pay increases was met by obstinate resistance from the unions, whose Friday strikes brought the government to its knees. Tindemans' second government (1977–1978) went back to using classical Keynesian methods, as a result of which the government itself created jobs, as it were, for almost 200,000 people. My first government (April 1979–March 1981) continued this policy, which became increasingly burdensome on state finances. There was a decrease in the number of employees in the private sector, and taxable income shrank as a result. In 1985, the level of taxation rose to 46% of the GNP in comparison to 23.8% in 1955.

This also resulted in a new wave of bankruptcies. Between 1975 and 1981 the average cost of unemployment rose from 1% to 10% of GNP. The stalemate was unsolvable.

By the beginning of 1981, all fundamental aspects of economy had gone off balance. Belgium was considered the pariah of Europe at the time. In October of 1980, Jacques de Larosière, Director General of the IMF, compared our country to a car driving in the dark and heading for a wall at high speed. It was in this climate of crisis that I launched my rescue plan in March 1981: a drop of 5% in income (including benefits) and a suspension of the index till December 1981. The plan met with resistance from the Socialist ministers and my government collapsed. My Christian Democrat successor, Mark Eyskens, formed a weak government that lasted barely six months (March–September 1981). It was renowned for its excessive budgetary deficit of 12.9% of GNP. The late Guy Mathot, a Socialist politician who had managed to become Vice-Premier and Minister of the Budget at an early age, uttered that famous phrase, comparing the national debt to a common cold: "It comes on and disappears of its own accord". But it was not the budget that brought the government down; it was steel. The Walloon Socialists demanded guaranteed subsidies for their languishing steel industry and refused to participate in the Council of Ministers. Mark Eyskens criticised his "striking ministers" in public, and it was during this sorry state of affairs that new elections had to be called.

Give Me the Liberals Any Day

In the general elections of 8 November 1981 our party suffered its worst defeat since the Second World War. Its share of the Flemish vote went down by seven percentage points. The result for the French-speaking Christian Democrats was a complete disaster. The Christian Democrats consequently found themselves on the same electoral footing as the Socialists. The victors were the Liberals. My party needed time to recover from its defeat. Our activists were overcome by bitterness and indecision. They could not understand why the voters had punished our party so heavily and had let our coalition partner go virtually scot-free. Our ministers were criticised for their lack of clear profile and for their leniency towards the Socialists.

The voters had not made the solution to the crisis any easier. Having at first refused, I accepted the role of *formateur* on 7 December. Leo Tindemans was also eligible for the task, but the King was of the opinion that a coalition government with the Liberals would have a greater chance of succeeding if it was led by someone also acceptable to the Christian trade unions. I offered Tindemans the post of Foreign Affairs. "Only for a few months I suppose!" he replied. "No", I said, "for eight years or possibly even ten". I kept my promise, for Leo Tindemans held the position of Minister of Foreign Affairs till he became Member of the European Parliament in June 1989.

Some left-wing journalists were furious with me for negotiating with the Liberals and questioned my motives for forming a "right-wing" government. But these were seasoned politicians with well-thought-out views who propagated a worthy form of socially tinted liberalism, a revelation to the Christian Democrats following all the paralysing action undertaken by the Socialists. We saw eye to eye on a policy for recovery. This *idem velle atque idem nolle* would continue to typify my fifth government and form the basis for its enormous decisiveness. I have never kept secret my preference for sitting at the negotiating table with the Liberals rather than the Socialists. Only very seldom did any embarrassing situations arise. With the Socialists the council of ministers regularly ended up in verbal brawls. I tried then to absorb the tension and not let the roughness of their threats injure me but over time it became all-consuming for me. It made me become more reserved.

During the negotiations for the new government I did not mention my plan to devalue the franc, because I knew that the chief Liberal negotiator, Willy De Clercq, would baulk at the idea and leave the negotiation table. The night of Saturday 12 December proved crucial, as I had reached an agreement to suspend the index. It still surprises me how easy it was to agree on this issue, given the fact that over the years no politician had had the courage to take such a decision. The government agreement provided for special powers for a period of one year, a strong recovery policy, a lowering of wages and a redistribution of income that would affect not only the employees but the self-employed as well. But not a word was mentioned of the devaluation of the Belgian franc. The government agreement was strikingly brief: a mere sixteen pages. "No more Bible, the government has to decide", proposed one of the party leaders.

I wished to gather together all the political "tenors" in my fifth cabinet. The Liberal leaders Jean Gol and Willy De Clercq and the leader of the French-speaking Christian Democrats, Charles-Ferdinand Nothomb, became Vice-Premiers. Of those selected from among the Flemish Christian Democrats, Leo Tindemans became Minister for Foreign Affairs, the outgoing Prime Minister Mark Eyskens was appointed to the Economy portfolio and my former principal private secretary, Jean-Luc Dehaene, was given Social Affairs, a position he held with verve. Jean-Luc was not the only stalwart of the Christian Workers Movement in the government. The movement was strongly present through other individuals as well, its task being to make this new regime of sobriety palatable to their members.

Devaluation

The situation had not improved during the hectic months of November and December 1981. There was still heavy speculation on the Belgian franc, the army of unemployed grew by the month, closures and bankruptcies followed each other at a brisk pace and the financial abyss the country was staring into grew deeper every day. My government knew from the very day they took office that they had to implement an emergency plan straight away. Otherwise the IMF would force humiliating conditions upon us if we wished to remain on top of the exchange rates. A huge amount of interest on loans hung like the sword of Damocles above our heads. Fortunately, the government had managed to remove quite a number of obstacles in its way from the outset. It had already separated itself from the social partners, who now could no longer determine socio-economic policy. "The government shall govern" was our motto. The special powers that allowed us to amend or abolish laws of our own accord without having to go through weeks of debate in Parliament caused even more of a stir.

Why did I choose to avoid conducting a full parliamentary debate? My experience as a convinced parliamentarian had proved a great disappointment: governmental work often got stranded there or ended up being blocked. Each bill had to be read at least four times: in Parliament and in the Senate, and before that in the various committees concerned. By the time a law could be voted on, this time-consuming procedure had emptied it of all its substance. The bills had long since been superseded by events and ended up being nothing more than empty shells. I had asked for special powers not out of fear of being rejected by Parliament but rather to be able to work more quickly and efficiently. How could I have pushed devaluation through if I had not managed to make a decision on postponing the index that very same weekend? Had I followed normal parliamentary procedures, the effect of devaluation would have been reduced to nothing. There was much ado about this afterwards. People were quick to call these powers "mandates", but that was not the case. Mandates allow the Parliament to be overruled, but in the case of special powers Parliament still holds sway over the government. The Parliament could stop them at any time and therefore retained its veto. In fact, I have never been in Parliament as often as then. I went there to explain every decision taken, following which a vote of confidence in the government was held. There was seldom as much debate as during that period. Debate in Parliament took up an awful lot of time but at least the cabinet got its work done and that was what mattered.

Many wondered whether the recovery programme would be enough to turn the tide. The National Bank stressed that devaluation could only be avoided if we followed a draconian regime comprising a 10% drop in income and prices. But such straightforward sacrifices were unfeasible. It was written in the stars that all this would end in devaluation. At that time, I still doubted the expediency and appropriateness of such a measure. For years it had been portrayed as something the people would never accept. No government would survive such a measure. Among the conservative financial circles in the National Bank in particular, devaluation was considered as simply *not done*. Heads of government who were unable to maintain the franc's exchange rate would be signing their own death sentence. This too made me doubt devaluation.

It was about that time that Fons Verplaetse walked into my life. He was a member of the think tank that had been analysing the prospect of devaluation since 1980. In December 1981, he became my deputy *chef de cabinet* and explained with much conviction that the increasing deficit in payments was our greatest problem. The deficit undermined our competitiveness and had to be tackled and repaired first. I could see only one solution for this: a devaluation of the Belgian franc. Together with my *chef de cabinet*, Jacques van Ypersele, later to become the *chef de cabinet* of the King, Verplaetse drafted a report, and I seized on the analysis it contained with open arms. The report provided irrefutable proof that the Belgian economy had progressively run up a 12% greater deficit compared with its neighbouring countries. It concluded that moderation alone would not suffice to re-establish the balance, as the deflationary effect was too strong and demanded forfeitures of 3%–4%. The rest of the handicap needed to be tackled by a devaluation of at least 8%. This would allow us to rid ourselves of part of our misery abroad, as Belgian products would suddenly become 8% cheaper. The other side of the coin, that is, inflation resulting from more expensive imports, could be compensated for by a suspension of the index and by calling a halt to price increases.

Verplaetse's arguments were so convincing that he won me over for good. It was then my task to convince my Liberal Minister of Finance, Willy De Clercq. Like the executive of the National Bank, he feared that this devaluation would have "no bite". If other measures, such as suspension of the index, were lacking as a result of pressure from the unions, the country would be in danger of being caught in a downward spiral of inflation and devaluations. On 14 February 1982, I met Willy De Clercq in our hometown, Ghent. There we took one of the most difficult decisions of our political careers: we decided to devalue the franc that following weekend and give Belgian exports a considerable boost as a result. But, before we got that far we still had to inform one particular man about our plans: Jacques de Larosière, Director General of the International Monetary Fund. And we had an ideal occasion for a meeting. As Prime Minister of Belgium, I was also President of the European Council for the first six months of 1982 and protocol required that I pay a visit to President Ronald Reagan. The visit had been scheduled for the week before devaluation, so I could also pay a visit to the head of the IMF in Washington at the same time.

In order to keep as many journalists out of Brussels as possible, we flew to Washington on a Belgian Air Force DC-10 with a sizeable delegation from the press. During my stay in the American capital I gave Jacques van Ypersele the task of arranging a top-secret meeting with de Larosière. I lodged at Blair House, the guest residence for visiting heads of state and government. Jacques van Ypersele came and woke me in the middle of the night to inform me of the contacts he had made. We could visit the IMF Director General at his apartment, the only problem being discretion. We were surrounded by very attentive security guards led by a highly enterprising woman whose sharp eye never missed a detail. During dinner at the Belgian embassy on the evening before our morning appointment with the IMF, van Ypersele whispered in the woman's ear and explained that the Prime Minister wished to go shopping alone the next morning at eight O'clock without being surrounded by bodyguards. She pointed out that the shops in Washington would be closed at that early hour, but she seemed to understand my problem. Did I wish to have an intimate meeting with someone? That could be arranged. In desperation, my *chef de cabinet* told her the true story and agreed to a security escort for the long drive to de Larosière's apartment.

On the morning of 18 February, we sneaked out of Blair House through the staff entrance and made our way past the garbage bins. When we arrived at de Larosière's residence four security guards were already waiting for us. Following a talk on the accompanying measures, he recommended a devaluation of more than 10%. This ultimately became 8.5%. The decision was taken in principle and the secret was well kept. Now it was a matter of putting this delicate operation into practise. Once back in Brussels, I called a meeting of the core cabinet on Friday, 19 February, followed by a meeting of the Council of Ministers at 11 a.m. I noticed that some of the ministers were visibly trembling. When I later met the Governor of the National Bank together with Willy De Clercq, the encounter proved dramatic. The Governor and his successor, who would effectively take over from him a few weeks later, were totally opposed to our decision. The news hit home like a sledgehammer. When I asked the Governor to attend a meeting of the European Monetary Committee (a meeting of experts preceding decisions taken by the EU Ministers of Finance) the following day, he categorically refused. I was forced to use my authority. He was pale as a ghost, and only after much insistence did he declare he was prepared to attend.

Immediately afterwards, I informed Pierre Werner, my counterpart in Luxembourg. Up until that moment, we had not involved the government of the Grand Duchy in the secret negotiations, even though it was their currency too. I knew that this was not correct but I greatly feared indiscretions. The Luxembourg government was renowned for being unable to keep files secret from diligent financial news reporters. Jacques van Ypersele told me we could afford no risks because if the plan were leaked it would increase speculation on the franc, which would end in a national disaster.

Nor were we thanked for our secrecy. Pierre Werner was particularly piqued. His country was being dragged into an operation they had no need of, as Luxembourg's public finances were in a healthy state. I told him that we were highly embarrassed by this incorrect way of conducting affairs but that we had no other choice. Fortunately, I was able to regain Pierre Werner's trust. I pushed a decision through to pay compensation for a Luxembourg TV mast that had been destroyed by a Belgian fighter jet. Without giving any good reasons for the delay, our Department of Foreign Affairs had allowed the case to drag on for years.

A Battle for the Future

As the English saying goes, "A baby is easy to conceive but difficult to deliver". This became obvious not only from the reactions of Pierre Werner but also from the recalcitrance of our EU partners. As is common during sound negotiations, the Belgian delegation proposed a higher percentage than they expected to be given, that is, 12%. But no one seemed prepared to grant us this competitive advantage. Karl Otto Pöhl, the Bundesbank President, said he would allow only a 3% devaluation. Our enraged Luxembourg partner brought in the heavy artillery in an effort to block our demand.

Negotiations were rendered even more difficult when news of the demand for devaluation leaked out. The British press agency Reuters published the news. I was in a meeting with my chief ministers at Val-Duchesse at the time. Tension had reached a peak. A throng of reporters had gathered outside the castle gate waiting for an official statement from the government. But because the Monetary Committee was still discussing the matter we were not yet planning to let the world know. A compromise was finally reached on Sunday evening of 21 February. After hours of arguing with his European colleagues, Willy De Clercq managed to salvage a devaluation of 8.5%. Once he accepted a task he committed himself to it completely. That was typical of Willy. Great was the relief when we were given this liberating news. We immediately took a decision that countered the National Bank's pessimism. Using our special powers, we postponed the indexed increase on wages and benefits. We left the minimum wage untouched. We also announced a three-month freeze on prices. This would compensate for the pressure of inflation, thereby optimising the effect of devaluation.

Now, twenty five years on, I can state that devaluation did bring about the long-awaited recovery of the economy. But it did not have a positive effect only on the economy. It also had a deeper psychological significance. The government pushed drastic measures through on its own and freed itself from the grasp of its social partners. In a single weekend we crossed the Rubicon armed with a legion of decisions that brought about a definitive turnaround. The unions were so perplexed by our will to act that they implicitly accepted our actions and, as a result, did not mobilise their members against us. The insurrection that had been predicted among the people for so long failed to materialise. Even the socialist union was forced to accept this unavoidable fact.

The question remained: could the population bear a further decrease in spending power? They had long since forgotten what it was like to tighten their belts, and even during those years of crisis they had continued to live their carefree exuberant "Burgundian" lives. Because the unions found themselves with their backs against the wall, they could do nothing but accept devaluation. A few months later, however, the psychological effect of the measures had ebbed away and impatience about the expected results mounted visibly. We would only have a chance of succeeding if we could convince the unions to tolerate our policy and not fight against us. But that was more easily said than done. There was a real risk that the two unions would join forces, forming a common opposition front and setting the country alight with massive demonstrations and lightning strikes. The Socialists were ready for action. But the Christian union kept an open mind. Their leader in particular, Jef Houthuys, was convinced. He agreed with the motto "Short-term pain for long-term gain". I knew that he was willing, if it proved necessary, to push though a socially acceptable policy even against the wishes of his supporters. More than anyone else he recognised that indifference and lack of foresight were real dangers for our society.

We wanted to pull the country through this bottleneck, but we tried at all costs to prevent anxiety and sorrow among "the little ones", as we called them. Jef did his utmost to distribute the load and consequences in such a way that the heavier burdens fell on the shoulders of the strong. Thanks to his efforts, a postponing of the index took the form of "cents instead of percents", as a result of which those in the lower income bracket remained protected. I thrive on and feel at home at informal talks like those held in the Ardennes at the time. Despite all logical systems, government bodies and negotiation mechanisms, I always come back to people. Whenever I reflect on it, I realise that my life in politics has always gone further than just putting one's trust in others. It is also a matter of human friendship and love. This is a wonder and even more than a wonder. Yes, love. People who are not familiar with the Christian Workers Movement might find this hard to understand. We do not call each other comrade but friend. We can argue and argue and point out each other's mistakes and failings. But we can also choose to put them aside and forget them. We can forgive.

I often asked myself if we did enough to save the state's finances. If the state goes bankrupt, so do health insurance, pensions and unemployment benefits. And then the common person is always the first to suffer. Allowing the national debt to swell is an outrage against society. The interest has to be coughed up by the average employee and by those with lower incomes, and it all returns to a select group of those who invest in state bonds. But, as Fons argued, "if we regain competitiveness, state revenue will increase by itself and the debt will melt away like snow in the sun". And he added, "You have to choose. You cannot do two things at the same time, that is, both restore competitiveness and heal the state's finances". I instinctively revolted against this reasoning and, unfortunately, I was proven right. The debt did not melt away like snow in the sun; it brought about an interest avalanche.

No Turnaround

On 13 October 1985, I had an important appointment with the electorate. Pundits felt that we would be heavily punished. "No turning back" was the slogan we began our campaign with. It was highly unusual to go to the polls with such a clear, direct message. Those who voted for us knew beforehand that they were voting for the same coalition and the same policies. There was a broad consensus among Christian Democrats; everyone realised that alternatives that were not serious enough would end up having dramatic consequences. The Socialists would be left standing in the wings and that proved to be the right choice. One had only to read their programme to realise that they had learned nothing. They did not even accept the recommendations made by international organisations. On the contrary, they proposed that the whole reform process should be reversed. And this would have led instantly to the same catastrophe experienced in 1981.

I became the figurehead of the campaign but was advised not to take the whole burden on my shoulders because if it failed, I would be forced to foot the whole bill personally. But I did not doubt for a minute. The 1981 defeat had taught us that we needed more than just a good message with substance. The message had to be personified by someone who embodied the message. I attached a lot of importance to honest communication with the voters and told them that there was no room for a change of heart; we would need to hold on for another two years to bring the economy back to health. When the results began to come in on Sunday evening, 13 October, I experienced a moment of happiness: against all expectations the government emerged from the battle even stronger than before! The results showed that the people were clearly willing to carry through the reforms; on the condition, of course, that politicians demonstrated vision and responsibility and could express it all in a coherent plan. Blockades of many years could be torn down by appeals to common sense. Helmut Schmidt in West Germany, Raymond Barre in France and Margaret Thatcher in the United Kingdom had realised this much earlier. The change was now occurring in Belgium. There was a sigh of relief in European circles and congratulations were sent. Thanks to the willingness of many who could put up with a decrease in their standard of living, we were freed from this economic swamp.

Guy Verhofstadt

The message sent by the voters was so clear that the King kept his consultations brief. I was appointed *formateur* on 16 October. I set to work forming another Christian Democrat–Liberal cabinet. My decision was the result of an agreement I had made on election night. I had made this agreement with a crestfallen Guy Verhofstadt, for the Flemish Liberals of whom he was the leader were the only government party to receive a beating at the polls. However, he was not the type of man to lower the flag after a defeat. I wanted to reduce the national deficit to 7% of GNP by the end of our term in office. In 1985 it was 11%. By way of comparison, the average national deficit in the other European countries at the time was 4.7%.

So Belgium still had a long way to go. In principle, a budget in deficit can be balanced either through increases in revenue or through reductions in expenditure. Martens V opted for a sensible combination of both.

Guy Verhofstadt believed that this would not work. "Real restructuring", suggested the flamboyant neo-Liberal, "is cutting spending by introducing radical structural measures". This is why he threw his proposal so crudely onto the table: his party could accept the government agreement only on condition that a fiscal moratorium be imposed. So there would be no new extra revenue! I watched him closely throughout the negotiations and tried to grasp his ideological premises and discover why he was waging such a bitter war against "the welfare state", as he called Belgium. For forty days he raged on about how a halt had to be called once and for all to all the waste in social security, to an unfocused, profligate administration and to the quasi-Marxist fiscal pressure on employment. We were unaware at the time that he would continue to repeat this line, harping incessantly in an almost Thatcherite fashion, practically oblivious to the opinions of his colleagues, who would rather have managed the budget by more classical means. I let him continue. I have always given my support freely to politicians with the gift and courage to plead for reform. A political class no longer moved by a project or a set of ideas is reduced to a mere management machine that passes the odd law and governs somewhat. How many politicians are driven by some higher project, a certain movement in thought, by outrage or by a vocation?

In the spring of 1986, I tried in a single move to push the budget in the right direction. We had locked ourselves up in the Château Saint Anne near Val-Duchesse. On the night of 19 May, we were able to put the finishing touches to our savings plan. It comprised a long list of measures designed to reduce the deficit by 4.9 billion Euros by the end of 1987. But my government were taken to task on the matter. In the wake of the Saint Anne round, strikes broke out in the public service sector, state companies and in education. The staff of Catholic schools also took to the streets. I heard from Jef Houthuys that, according to his executive, the burden had come down too heavily on the workers and those dependent on public benefits, whereas tax fraud had been left untouched. As always, the union leader stood bravely in the line of fire but, despite all his efforts, he was unable to prevent his members from taking part in the wave of strikes. It was not only between the Christian union and the government that tensions were visibly mounting. There were continuous conflicts within the government itself and Guy Verhofstadt's position was also an obvious cause. Some party members reproached me for sharing his enthusiasm for the process of atonement we wished to subject the floundering state apparatus to. This sounded the death knell for unity in the government.

In 1981 the budgetary shortfall in Belgium had reached 13% of GNP. Ten years later it was 6%. My government halted the snowball effect, but even that was not handed to us on a platter. The people had to transfer a part of their income and purchasing power to the government and to the enterprises. Following the collapse of my government in October 1987, a myth about Verhofstadt began to take shape and circulate. He began to be presented by his supporters as the wonder boy of the budget, the one who had managed in just two years' time to do all the essential work involved. This is a gross exaggeration if we examine this restructuring in the longer term. Propelled by Philippe Maystadt, the Christian Democratic Minister for the Budget, government expenditure dropped 6.1% points during the time the centre-right cabinet was in office from 1981 to 1985. The Liberals' contribution to this decrease was limited. From 1985 to 1987, during Martens–Verhofstadt, there was a further decrease of 2.6% points and from 1988 to 1992, during my last term in office, it dropped again by another 2.6% points. Thus, it is hard to call this the Verhofstadt effect. The greatest efforts I made date back to the time when he was not even a minister. Yet, if Verhofstadt and I had managed to complete a full four-year term in government, things would certainly have turned out differently.

My personal affection for Verhofstadt stems from the fact that he was a fresh Liberal thinker who charmed me with his constant flow of suggestions. Even though he was my fiercest opponent in my own constituency, we still got on well together. There was a sense of cooperation between us that can only exist between people from Ghent. At that time observers insisted on calling him my political son, whereas I termed him "the Mozart of Belgian politics". But we parted ways painfully in May 1988 when I accepted the position of Prime Minister of the new coalition Jean-Luc Dehaene had put together between the Christian Democrats and the Socialists, now almost twenty years ago. For eleven years Guy Verhofstadt fought a bitter fight in the opposition. But he reformed the Flemish Liberal Party and won the 1999 elections. For nine years he was Prime Minister of a "purple" coalition, composed of Liberals and Socialists. This was possible because, as government leader, he transformed himself into a highly pragmatic politician who reconciled the Liberals and the Socialists. For the first four years, he even brought the Greens into his government. His ambition was to accomplish three major objectives: to transform Belgium into a model state, to reduce government taxation drastically and to build up considerable state reserves to be able to face the demographic ageing of society. He failed to push any of these objectives through Parliament, let alone begin to carry them out. The voters found his policies incredible. In 2007 Belgium's tax burden was one of the highest in Europe, 46% of GNP. He was presented the bill at the general elections of 10 June 2007. It turned out that Verhofstadt was no wonder boy after all and that the truth about him is to be found somewhere between the two extremes.

The King Refuses to Sign

In late March 1990, my last government was faced with an unprecedented constitutional crisis. King Baudouin refused to give royal assent to a law on abortion passed by Parliament. This placed my party, which, like the King, was also opposed to the law, in a very delicate situation. Although I had helped determine our party's position on the matter as a young party leader, up until that moment the abortion issue had remained dormant. In all of the negotiations that had been held since that time, it became clear that no compromise could be reached between those in favour of far-reaching liberalisation and those who believed that abortion could be legal only in exceptional, well-defined circumstances. In their final report issued on 15 June 1976, the twenty five members of the State Commission on Ethical Problems declared in favour of a change in the law. A narrow majority of thirteen mainly free-thinking members stated that they were in favour of abortion being removed from the penal code, and cited in this regard considerations of a socio-psychological nature concerning women in need. The minority of mainly Christian Democrat members were only in favour of removing medical abortion from the penal code, thereby still requiring that reasons of health be objectively demonstrated.

In the years that followed, dozens of bills had been submitted but no majority decision could be reached on any occasion despite the fact that public opinion had become increasingly tolerant on the issue. The decision by the Public Prosecutor no longer to prosecute abortion was met with a high level of agreement among the population. In April 1986, two senators, a Socialist and a Liberal, tabled the umpteenth proposal: abortion could be carried out at a woman's request during the first twelve weeks of pregnancy; in the period that followed abortion could only be carried out in special circumstances and under care. In November 1989, after more than three years of debate, the Senate accepted this bill.

As head of government, I had remained outside of the debate. Yet everyone knew my position on the matter. I had declared that I was in complete agreement with the position of the Belgian bishops, according to whom the bill was a radical denial of the right to life in its early stages. Following its approval in the Senate, I encouraged my party to table their own bill and made efforts to reach a last-minute compromise. Numerous formulations were tried and tested in an effort to reach agreement, but the Socialists did not trust us and resisted all of our proposals.

Though the King had never discussed the matter with me, there was doubt in political circles that he would ever sign a law on abortion. This fear became real when he stated in no uncertain terms that the right to life before birth was an important matter to him. In this respect he referred to the internationally adopted 1959 Declaration of the Rights of the Child, which states that "the child, by reason of his physical and mental immaturity, needs special safeguards and care, including appropriate legal protection, before as well as after birth".

On 30 March 1990, the Parliament also passed the bill. My surprise was therefore great and sincere when, upon being invited to an audience with the King on the day following the passing of the act, he presented to me the draft of a letter declaring that his conscience would not allow him to sign and thereby to give assent to the law on abortion passed by Parliament. I read and reread the draft in the company of the King. In the blink of an eye, I could see all the possible ramifications of his refusal. Whether it was the muse or the Holy Spirit that inspired me, I do not know, but I said, "Sire, your letter sets out a fundamental problem. As the letter stands, I can do nothing else but hand in my government's resignation. This could lead to a constitutional crisis and a crisis involving the monarchy. If you wish to avoid this, you will have to request me to find a solution combining these two factors: your crisis of conscience and the good functioning of our democratic institutions". The King took back his draft and had a letter sent to me a few hours later. Much to my relief, I noticed that the King had appended a request that a legal solution be found for the problem.

A Difficult Quest

The quest for a solution could now begin. The talks had to be held in the utmost secrecy. A leak to the press could have brought about an unprecedented political crisis. I called my Vice-Premiers to a meeting on the Saturday evening to exchange initial ideas on the matter. It turned into a brainstorming session that explored all possible outcomes, even the most radical. Most of these possibilities were effectively set aside, were shown to be unfeasible or were rejected by the Palace. I continued to insist on absolute secrecy. The Parliament and the outside world could only be informed once a feasible solution had been found. In the meantime, each of the Vice-Premiers had visited the King privately. Their individual reports confirmed his intransigence on the matter. The King had stated the following both to me and to a few Vice-Premiers: "You can send the cardinal or even the Pope to see me tomorrow; I will not change my position".

After we had adjourned without a solution that Saturday evening and the following Sunday, I became even more aware that only the use of Article 93 of the Constitution could render a quick and discrete solution possible. This article states: "Should the king find himself in a situation that makes it impossible to rule, the ministers, after having established that this situation exists, shall call an assembly of both chambers. Guardianship and regency shall be provided for by the united chambers".

One first reading, this article seems intended only for cases in which the king is severely physically or mentally handicapped, but it also brought to mind several instances in which, according to the Prime Minister at the time, King Leopold III, then living in Brussels under German occupation, had found himself in a situation that made it impossible to rule.

We also failed to reach a solution on Monday, 2 April, and a stalemate loomed. As a result, I presented my idea to the Vice-Premiers. During a personal audience, the King provided me his guaranteed agreement to the invocation of Article 93 of the Constitution.

Following this favourable exploration of the terrain, I wrote a letter to the King on 3 April 1990 stating that during the period of incapacity to rule, the King's constitutional powers would be exercised by the Council of Ministers under their own responsibility, and that I would propose to them that they ratify and promulgate the law on abortion. Following ratification, the Council of Ministers would propose to Parliament that the King resume his constitutional powers; deliberations would then take place among the united chambers during which an end to the period of incapacity to rule would be established. The King answered only an hour later, stating that he agreed to the invocation of Article 93 of the Constitution.

With Beating Heart

My heart beating, I waited for the meeting with the ministers at Stuyvenberg castle that evening. We still had to agree on the matter of the King stepping aside temporarily and on their signing the law in his stead. I was all too aware of the historic importance of the moment and of the shock that would spread throughout the land once the media were informed of our solution.

The atmosphere at the meeting of the Council of Ministers was highly charged from the outset. I first read through the correspondence with the King, and then the ministers proceeded to reach agreement on the decision, accepting the fact of the incapacity of the King to rule. This decision was signed by each minister. Following this, the Council then decided to ratify and promulgate the bill on abortion. This law was signed by each minister present.

The die was cast on Wednesday, 4 April 1990. At half past five in the morning listeners heard for the first time that the King had refused to give royal assent to a law passed by Parliament, and that my government had declared him to be incapable of ruling. The news landed like a bombshell. Nobody seemed to grasp its real importance at first. Jurists and constitutionalists consulted on the spur of the moment set about improvising on the question. There was a great political stir. From Strasbourg, two Socialist MEPs openly called for King Baudouin's abdication. But as the day went on and spokespersons acquired the necessary information, a sense of levity ensued.

Calm and fully confident of a favourable outcome, I addressed the packed semicircle of Parliament and the bulging public galleries, first reading the King's letter, followed by my reply and then his response to the initiatives the government proposed to take. The meeting went off without a hitch. The incapacity to rule had lasted a mere thirty six hours, but almost everyone agreed that this was an exceptional occurrence and must remain so.

I look back with mixed feelings on those politically unsettled times. Though the King thanked me warmly and at length for the way in which I had managed to salvage the monarchy from this stalemate, I was given to understand from some of his comments that he was disappointed with the law on abortion because of the fact that no consensus could be found for an alternative, humane solution. I could understand his position and was essentially in complete agreement with him. Baudouin would certainly have abdicated had we allowed the situation to reach a crisis point. His most personal, intimate conviction was at stake. Personally, I was in favour of a limited amendment to the bill. If we had worked out a bill while there was time, as the Christian Democrats had done in Germany, we would now have more restrictive laws on the matter.

Chapter IV

Small Country, Large Responsibilities

It is sometimes said of Belgium that it is small inside but big outside. Indeed, for geographic and historic reasons, Belgian foreign policy is closely linked to its internal politics and vice versa. There is generally widespread consensus about the nature and degree of relations with other countries; sometimes foreign policy is internalised, however, and the classic Belgian differences of opinion crop up in dossiers that at first glance have little connection with national politics.

Since the Second World War, the main areas of focus for Belgium's foreign policy, besides the European integration process, have been transatlantic cooperation, especially in the context of NATO, and relationships with its ex-colonies – Congo, Rwanda and Burundi. In each of these three areas, I was confronted as Prime Minister with serious crises, and each time I played an active role in stabilising, improving and intensifying the external relationships. This was pre-eminently the case during the cruise missile issue, which put pressure on our good relations with the United States; the troubled relations with Mobutu in what was then still called Zaire; the ethnic tensions in Rwanda that preceded the genocide of 1994; and the re-launch of the drive to European integration after a long period of "Eurosclerosis".

Rekindling the Cold War

Between 1947 and 1989, international events were governed by the Cold War, a global conflict that divided the world into two competing power blocks under the hegemony of Moscow and Washington. The rivalry between East and West led to an unprecedented arms race. On each side of the Iron Curtain, a gigantic arsenal of nuclear weapons was amassed, capable of annihilating the earth and threatening the human race with a nuclear holocaust. Even though the deployment of these weapons of mass destruction became unthinkable in time, the countries of NATO and the Warsaw Pact were permanently on standby for the ultimate confrontation, which everyone hoped would never happen. The psychosis of fear engendered by this has never been equalled in history.

For anyone who grew up after the collapse of the Soviet block in 1989, it is difficult to have any idea of the existential fears that were raised by the prospect of nuclear doom. When I look at my youngest children, it sometimes occurs to me that they are almost as far away from the Cold War as I myself was from the First World War when I was the same age. And what did the Great War mean to me? A conflict with a rather archaic feel to it, which certainly appealed to my imagination when I was a young boy, but in my mind it always belonged more to the nineteenth than to the twentieth century. I never cease to wonder at the speed with which our world changes; events that today still arouse passionate reactions will, by tomorrow, have been reduced to a dull paragraph in the annals of history. The commotion caused by the building of the Berlin Wall, the Prague Spring and the Vietnam crisis now belong to past history, and the massed outcry of the Peace Movement against atomic weapons has long since died away in the streets of the European capitals.

When my generation entered politics at the beginning of the 1970s, the Cold War had been underway for twenty years or so. We regarded the "warm" phase of the East–West conflict as definitely past and peaceful coexistence and détente as an irreversible fact that international policy makers were duty-bound to pursue and continue. So it was a great disappointment when, at the end of the 1970s, attempts at a rapprochement between East and West came to a dead end and military confrontation flared up again. The United States, still burdened by the trauma of Vietnam, reacted particularly fiercely to the Soviet invasion of Afghanistan in December 1979. Cold War rhetoric intensified, both in Washington and in Moscow. The arms race switched into high gear.

Security issues also took on a new urgency for the European allies when the Soviet Union began unilaterally to modernise its arsenal of missiles in Eastern Europe. Starting in 1977, it replaced its old SS-4 and SS-5 missiles there with very accurate ground-to-ground, mobile SS-20 missiles armed with three warheads each. The few intermediate-range ballistic missiles that NATO could counter with were out of date and did not provide adequate defence against the threat emanating from the Russian SS-20s. According to NATO strategists, this gave rise to a dangerous imbalance of power, which might well tempt the Soviets to consider a limited nuclear attack on Western Europe or at least to threaten it with one. It was, therefore, more important than ever that the Western European allies could count on the American nuclear umbrella to protect them.

However, since the United States and the Soviet Union had reached parity in their intercontinental nuclear arsenals through the SALT agreements, uncertainty had arisen in Europe about the scope of the American guarantee. Would the United States really run the risk of a nuclear reprisal in order to defend the West in the event of a possible limited Soviet attack on Rotterdam or Hamburg? This was the question German Chancellor Helmut Schmidt asked in October 1977 in an address to the International Institute for Strategic Studies in London. Schmidt delivered sharp criticisms of the Americans who, he believed, had overlooked European interests in the SALT II negotiations. Furthermore, he pressed for the unequal military situation in Europe to be rectified in such a way that the strategy of deterrence with respect to the Warsaw Pact would still be maintained.

The Americans' initial reaction was to dismiss Schmidt's criticism, in the belief that they had adequate nuclear weapons in and around Europe. However, 1979 brought a change in this. During an informal summit of the United States, the United Kingdom and West Germany, a twin-track policy was devised for NATO. The Americans would negotiate with the Soviets over the dismantling of the SS-20 missiles. If the negotiations failed, the alliance would then start to modernise its intermediate-range ballistic missiles in Europe. On 12 December 1979, this policy was ratified by the NATO allies, including Belgium, in the famous "twin-track" decision.

From a technical point of view, with the Pershing II–type ballistic missiles developed and produced by the United States and its extremely sophisticated ground launched cruise missiles (GLCMs) the Western alliance had the necessary tools to withstand the threat from the Warsaw Pact. Whether from a political viewpoint it would ever come to deploying these proposed new air-launched nuclear weapons in the European NATO countries was another, very tricky, matter.

When I became Prime Minister in April 1979, the international debate on the missile issue was still at its height. Schmidt insisted that a number of other NATO countries besides West Germany should station the new intermediate-range ballistic missiles on their territory. He – quite rightly – wanted to share this nuclear responsibility with other European allies. By June 1979, he had already spoken to me about it during one of my first European summit meetings.

As a result of a combination of military-related and political factors, five NATO partners – the UK, West Germany, Italy, the Netherlands and also Belgium, which had housed NATO headquarters since 1966 – were called on to demonstrate this solidarity. This meant that in a year when it had a large number of other issues to deal with internally, my government had to agree to the stationing of American nuclear missiles on its territory. The missile dossier was to cause serious turmoil during my periods in government, for almost five years. Crisis threatened several times. Even in my own party, the Flemish Christian Democrats, the stationing of the missiles caused passionate discussions and there were differences in public opinion, the likes of which had seldom been seen.

Ethical Considerations

Although I witnessed the birth of the Belgian missile problem as an exceptionally intense political discussion, in the initial stages I did not yet see it as a universal question of conscience that would deeply polarise our society. Gradually, however, I began to give intense thought to the ethical aspects of stationing atomic weapons. And then a number of additional questions arose. Did the new nuclear weapons make a nuclear war more or less likely? Was nuclear deterrence ethically justified? Were the missiles reconcilable with the official aim of NATO, to promote détente and to preserve a Western world without war?

I believe the arguments both for and against could be defended with dignity. The fact that I myself ultimately agreed to the deployment of the NATO missiles was a decision that formed part of the strategy of the twin-track decision, to make possible the eventual dismantling of all nuclear weapons in Eastern and Western Europe. I therefore shared the same ethical concerns as the Peace Movement. I also wanted the Berlin Wall to fall, the Soviet dictatorship to disappear and our continent to be freed from the nuclear stranglehold. But the key question was how and by what means we could arrive at effective results with respect to disarmament and relieving tension in East–West relations.

I was by no means in agreement with the methods of the Peace Movement, that is, unilateral disarmament of the allies and setting up a nuclear-weapon-free zone in Western Europe. The imbalance of power resulting from this would undermine mutual deterrence and, for the Warsaw Pact, lower the threshold for a mutually destructive nuclear conflict. In the interests of world peace, NATO needed to restore the military balance in Europe by modernising its intermediate-range ballistic missiles there. In that respect, the first section of the twin-track decision was morally acceptable.

Maintaining the balance in nuclear deterrence could not, in my opinion, be a goal in itself, however, but simply a step on the way to gradual mutual disarmament. To speed up that process, additional peace efforts were needed. Which is why, parallel to the decision to modernise, the Belgian government insisted that the Americans invite the Soviet Union to new negotiations. We found a great ally in Helmut Schmidt. At the NATO talks on 12 December 1979, he successfully got the offer to negotiate promoted to a full second section in the twin-track decision. The last sentence of the NATO undertaking read as follows: "NATO's TNF [Theatre Nuclear Forces] requirements will be examined in the light of concrete results reached through negotiations". I believe this clause contained the moral justification for our support for the twin-track decision. It not only strengthened West European defence, it also opened a window to negotiations and, therefore, to the control and reduction of weapons.

The fact that the negotiations that started in 1981 ran aground two years later cannot be ascribed to a lack of will on the part of the NATO allies. The zero option that NATO took as its starting point was, after all, a reasonable basis for negotiation in every respect. If the Soviet Union would dismantle the SS-20 missiles it had installed, there would be no new American missiles in Europe.

After the failure of the Geneva negotiations, many opinion makers believed that the chances of a disarmament agreement had been lost forever, especially once the NATO partners began their planned deployment of the nuclear cruise missiles. However, they failed to appreciate the favourable side effect that a temporary rearmament would bring. The revived arms race in fact brought the Soviet economy, exhausted by excessive expenditure on armaments, permanently to its knees and drove its leaders to the negotiating table of their own accord. For the Soviet leaders were in dire need of disarmament to dig their country out of an economic hole. Thanks to the goodwill of the young, dynamic Mikhail Gorbachov, in consultation with President Reagan, the zero option was indeed reached at the end of 1987. John Lewis Gaddis writes as follows in *The Cold War* (Gaddis, 2007, pp. 222–223):

> The two men never agreed formally to abolish nuclear weapons, nor did missile defense come anywhere close to feasibility during their years in office. But at their third summit in Washington in December, 1987, they did sign a treaty providing for the dismantling of all intermediate-range nuclear missiles in Europe. "Dovorey no provorey," Reagan insisted at the signing ceremony, exhausting his knowledge of the Russian language: "Trust but verify". "You repeat that at every meeting", Gorbachov laughed. "I like it", Reagan admitted. Soon Soviet and American observers were witnessing the actual destruction of the SS-20, Pershing II, and cruise missiles that had revived Cold War tensions only a few years before – and pocketing the pieces as souvenirs. If by no means "impotent", certain categories of nuclear weapons had surely become "obsolete". It was Reagan, more than anyone else, who made that happen.

Does this demonstrate that the strategy of the twin-track decision was the right one? Undoubtedly. By first placing more nuclear weapons in Europe, NATO was ultimately able to remove them all. If we had not taken the decision to install them, history would definitely have taken a different turn.

The apparently logical outcome of the entire missile problem in 1987 conceals the scrupulous political debate that preceded it. I am still surprised at the doggedness with which organisations and politicians opposed the twin-track decision, even though implementing it led to the zero option.

Local and Socialist Opposition

In December 1979, my coalition of Christian Democrats and Socialists was supposed to endorse the NATO modernisation plan. The bipartite line of thought concerning the twin-track decision fit perfectly with the coalition agreement and with the so-called 'Harmel doctrine' (named after our former Minister of Foreign Affairs, Pierre Harmel), which linked a strong defence to good diplomatic relationships with the countries of the Warsaw Pact. But that did not take into account the Flemish Socialists. In the autumn of 1979, they began to cast fundamental doubts on the cooperative stance of our country in NATO defence. Since I was leading a coalition government and needed the support of all the parties, this opposition was tantamount to a cabinet crisis. The Flemish Socialists demanded that the government insist on a moratorium of six months and, if we were alone in questioning the proposal, Belgium would just have to force an adjournment of the NATO Council. Belgium was alone in its reservation, since the Netherlands – which also opposed the deployment – did not have the courage to come to our country's aid. The Flemish Socialists would therefore have to abandon their unrealistic opposition. I did not intend to give in to their ridiculous arguments.

On Wednesday, 12 December, I summoned the President of the Flemish Socialists, Karel Van Miert, to my office. There was some urgency because a few hours later the Minister of Foreign Affairs, Henri Simonet, was to defend our position at the NATO Council. I attempted to make it clear to Van Miert that the moratorium no longer stood a chance. Seldom have I seen anyone so distraught as Karel Van Miert on that particular morning. This man who was normally always so animated was now painfully torn between his intuitive preference to frustrate the twin-track decision and his sense of responsibility to avoid a government crisis. I was often surprised during these turbulent days just how far he had distanced himself, with his viewpoint, from French President François Mitterrand and German Chancellor Helmut Schmidt, Socialists who supported the NATO strategy.

In 1989 Van Miert exchanged national politics for an appointment as a member of the European Commission, responsible for transport and consumer protection and, from 1993 to 1999, for European competition policy. As quite frequently happens with politicians who accept higher responsibility, he underwent a complete metamorphosis. The caterpillar became a butterfly, but immediately forgot that it had ever been a caterpillar. The route he has since taken has been of benefit to the liberalisation of previously monopolistic government sectors. The impassioned planned economist of yore now became, believe it or not, the fiercest advocate for a full-competition policy. He has rebuked many a European government for implementing business subsidies that interfered with competition. He has calmly reprimanded even the largest corporations on merger plans tending towards monopoly.

I was disappointed that the Flemish Socialists had so unexpectedly become the self-appointed figureheads of the anti-missile opposition. Fortunately, state interest ultimately gained precedence over hidden electoral strategies. A great deal of the credit for this goes to my Minister of Economic Affairs, Willy Claes, a Flemish Socialist and later Secretary-General of NATO. He liked to regard himself as one of the politicians who had the interests of the state at heart. It was decided that the Socialist Party would not leave the government if it approved the twin-track decision, on condition that a number of qualifications would be linked to the Belgian agreement, including the way the decision was implemented. Belgium at the same time expressly promised to stand by the collective decision of the alliance and endorsed the twin-track decision to deploy 572 American intermediate-range missiles in five Western European countries.

Now, so many years later, I still think it was the right decision. If we as the government of a small country had not agreed to the deployment at that time, we would immediately have squandered all our influence in the alliance and in Western international politics. There is no way we would still have been able to invest in the second pillar of the twin-track decision: negotiation. Any other stance would have undermined the cohesion and credibility of NATO, which would have given the Soviets the incentive to take an even more rigid line. Moreover, an abstention by Belgium would have badly damaged our international prestige and our interests.

Throughout all those years I saw the wave of protest against nuclear missiles in Belgium and the other countries involved steadily rise and develop into a mass movement impossible to ignore. Intellectuals and clerics declared that the decision had been "provoked by a regime founded on force". I totally disagree with this statement. No matter what one thinks about the way America behaves in the world, the country is an open and democratic society. In the United States I met Nobel Prize-winning scholars who openly encouraged me to postpone the deployment of the cruise missiles. That was quite possible there, without any reprisals, whereas the closed Soviet regime was rumoured to have been striving for ideological mastery of the world. In Geneva, the negotiators were not alike as far as their view of humanity or their control over their own political actions were concerned. The Soviet Union acted and negotiated passively, impenetrably and only to further its own interests. The United States at the same time had to inform, consult and unite the entire network of the alliance. Via NATO, they led their sovereign allies in free consensus.

Visit to Washington

The negotiations in Geneva began at the end of 1981, but when the Russians broke them off in November 1983 our country seemed unable any longer to escape from carrying out its commitment. When the infrastructure at the Air Force base of Florennes, a small town in southern Belgium, was ready at the end of 1984, the moment of truth had arrived. In January 1985, I went to Washington with my Minister of Foreign Affairs, Leo Tindemans, to explain the viewpoint of the Belgian government to President Reagan. That government viewpoint left room for the outcome of the negotiations with the Soviets and stated that the precise timetable for the deployment would be established in consultation with the allies. A great deal of time had been spent polishing our speaking notes, so as to bring both pro-Americans and those who were "anti-missile", primarily within my own Christian Democratic Party, into line.

On Monday, 14 January, the Belgian delegation was received at the White House with much pomp and ceremony. After I had written a brief message in the guest book I was invited, along with Leo Tindemans, to a separate discussion in the Oval Office, where American presidents receive their important guests. The world-famous room seemed smaller to me in reality than photographs at any rate seem to suggest. I was invited to sit next to President Reagan, who was assisted on this occasion by the Secretary of Defense, Casper Weinberger, the Secretary of State, George Schultz, and the National Security Advisor, Robert McFarlane. The first contact was friendly and unforced, though I sensed a certain scepticism.

It was not the first time I had met "the Great Communicator". I had already made the acquaintance of Reagan in June 1982 at a summit of the G-7 in Versailles, where I had been invited as President of the European Council. President Reagan seemed to me an optimistic man who laughed easily and radiated a strong Christian conviction. I noticed that during technical discussions he was rather absent-minded, and if he did add anything to the conversation his sentences were punctuated by simple biblical expressions. This made him to some extent the odd one out at the G-7 summit, where the other world leaders were only too ready to expound on their economic theories and views. It was, incidentally, the period when Helmut Schmidt was giving ingenious exposés of the way in which a halt could be called to inflation and unemployment. Was it because the President did not know enough about these dossiers that he avoided discussions on policy? It probably was. But this did not detract from the fact that he intuitively had a good idea of what he wanted and set out towards his goal in a very straightforward manner. "No" meant "No" and "Yes" meant "Yes". This soon became clear to me that Monday.

During the conversation I was surprised by how quiet my Minister of Foreign Affairs, Leo Tindemans, was. He, the pro-American who normally happily entered into discussion with the Americans, did not intervene once. I assumed at the time that he was not too happy about our asking for a further delay. He evidently could not see the point of it and so did not defend it. As a result, I was the only one supporting it in the Oval Office. Having dragged out my speaking notes, I explained to the President that the Belgian government no longer harboured any reservations about the deployment of the first sixteen missiles, but that it still wanted to question its allies about a possible postponement of the agreed starting date. At the end of my exposition Reagan pulled out a crib sheet, hidden behind his wedding ring, and said: "In any case, the United States will vote 'no'!" The President did not in fact have any objection to a final consultation, but the US rejected any further postponement out of hand.

The President and his staff were most surprised by the fact that my government had apparently still not adopted a definite standpoint. For them this fundamental matter had been settled long ago. They assumed (and our diplomatic corps should have anticipated this) that the date for deployment was dependent only on technical factors, such as whether or not the required storage sites and infrastructure would be ready on time. Personally, I found it incomprehensible that the Americans did not know about the Belgian reservations relating to the deployment and therefore about the basic fact that a crucial political decision still had to be made before they would be able to fly the cruise missiles over to Florennes.

In no time at all I could see my negotiating position collapsing like a house of cards. Where I had thought the Americans would be grateful to us because I was committing Belgium to completing the deployment by 1987 at the latest, Reagan and Shultz considered this as self-evident for a country that had ratified the twin-track decision. Reagan told me, moreover, that he had documents in his possession in which a Belgian minister had given political agreement to the deployment of, for the moment, sixteen missiles in March 1985. So while I was putting forward the case for a delay, the President was demanding from me an explanation as to why I wanted to go back on this previously agreed commitment. I knew absolutely nothing about this. I had absolutely no idea which document and which minister the President was talking about. What further arguments could I now advance against his logical objections? I no longer had a leg to stand on.

Afterwards I gradually found out about the insidious decision-making process by which both the Ministry of Foreign Affairs and Ministry of Defence had committed our country to the timetable years ago. To my amazement I was forced to come to the conclusion that there had actually already been a date for deployment from the very start. I had never been informed of the existence of a top-secret diplomatic memorandum in which the Foreign Ministry had given the American government; not a technical commitment but in fact a full political one. In every respect, this was a crass example of how civil servants make politically far-reaching decisions without the approval or even the knowledge of the political leadership.

Allies ...

While the public debate about the timetable for deployment was still in full swing, Leo Tindemans travelled to the main European capitals to sound out the allies about the possibilities of Belgium delaying for a while or spreading out the deployment of the nuclear missiles. This trip did not produce the envisaged results. Tindemans was received everywhere with a "no", even by his Dutch counterpart, Hans van den Broek, in spite of the fact that Ruud Lubbers' government had given itself a delay of at least eighteen months.

Margaret Thatcher invited me, accompanied by Leo Tindemans, to a meeting on Saturday, 2 March 1985. It was a remarkable conversation and by no means the last I was to have with the "Iron Lady". Mrs Thatcher received us at the beautiful Chequers Court, since 1921 the country retreat of British prime ministers. She broached the missile issue in no uncertain terms: "I assured President Reagan that Martens and Tindemans were reliable and honest chaps who would do what was agreed. Please do not turn me into a liar".

Later in the day I was driving with Leo Tindemans to the more distant country estate of Lord Carrington, the British Secretary-General of NATO. When I raised the proposal with Peter Carrington, I could see a certain amount of incredulity on his face. He replied that the allies would respond very negatively to any decision by Belgium not to abide by its NATO commitments. If smaller Member States like Belgium no longer participated in the collective defence this would, according to Carrington, deeply undermine the strength of the NATO alliance.

I then realised that our anglophone allies would never understand why the Belgian government wanted to qualify its standpoint on missiles. Thatcher and Carrington had been born and bred politically in the typically majoritarian system wherein government leaders can make decisions quite quickly and straightforwardly, since they need only to convince their own grassroots support. I made the point that in Belgium I had to endure quite a lot of resistance within the coalition government and even within my own party, hence our reservations. Carrington turned a deaf ear to this. Perhaps he thought it was a sign of incredible weakness, but the fierce opposition was now actually a reality that I could not avoid.

The KGB Listens In

The "No" from Thatcher and Carrington was the ultimate confirmation of what we had already sensed during our visit to Washington. Belgium had no other alternative than to proceed with the deployment. In the laborious talks in the following days I noticed that a majority were gradually succumbing to the inevitable. The negative responses of the European allies to our appeals evidently had the desired effect and produced a kind of mental turnaround.

Shortly afterwards, Leo Tindemans travelled to Moscow to attend the funeral of the party leader, Konstantin Chernenko. At the same time, he presented there our final proposal: Belgium would very shortly decide on deployment, but was prepared to postpone implementation for two months if the Soviets would give the Geneva negotiations a chance.

Tindemans' démarche in Moscow ultimately came to nothing. He did not even succeed in being received by Andrei Gromyko, the Soviet Minister of Foreign Affairs. Fortunately, the Belgian delegation knew what to do about this. Towards the end of the visit, Tindemans rang me to ask when the government was going to decide on the cruise missiles. "In two or three days after your return", I said decisively. I had scarcely put down the receiver before Tindemans received his invitation. As predicted, the Russian espionage service, the KGB, had tapped our conversation and informed Gromyko at once. However, he haughtily rejected the Belgian disengagement proposal. After the unambiguous Russian *njet*, it was clear to everyone that no further arguments existed for postponing the initial deployment.

No More Waiting

The government now made the decision straight away. The Council of Ministers met on the eve of the parliamentary debate on 15 March 1985. In a short time we had agreed on the statement I was to read out the next day. In this I called our decision the most difficult since the Second World War. Just after eleven O'clock in the evening, I called the American ambassador to tell him we had agreed to the deployment of the sixteen cruise missiles. Leo Tindemans, meanwhile, had a memorandum drawn up in which the decision was formally ratified. On the question of who should sign the memorandum, he said, "It is better if the Prime Minister signs the memorandum, then it will have more authority". I then took the letter to the American embassy. The ambassador immediately forwarded our agreement to the Pentagon. Two aeroplanes had been standing by in the States for weeks: a C-141 transport plane on the American West coast with sixteen nuclear warheads on board, and a C-5 Galaxy aircraft on the East coast with propulsion engines, in other words, the actual missiles. These aircraft took off immediately, and in fact were already in the air when at three O'clock in the afternoon I read out the government statement in Parliament.

From a democratic point of view, flying over the cruise missiles was completely justified, since in Belgium this is one of the few domains where executive power takes precedence over legislative power. In just about all other domains the government can take action only with the approval of Parliament. But in matters of security and foreign policy ministers do not have to wait for the agreement of Parliament. Parliament could well have dismissed my government if a majority had disagreed with the deployment. But even then the decision would have remained legally in force until a new government had repealed it.

However, public opinion remained divided over the deployment and some people had great difficulty with the way in which the decision had been taken. It was, after all, very rare in Belgium for a decision to be made with such forcefulness. At first glance it was enormously unpopular, and it looked as if we would pay heavily for it electorally. But contrary to all expectations, seven months later the Christian Democrats gained a resounding election victory. Political courage does have its rewards after all.

I can live with the responsibility I took on 14 March 1985. A different stance would have had serious consequences. The Belgians would have ended up in a different country, a country that no longer housed the headquarters of NATO. In that case, Brussels would also probably not have developed into the European capital it is today. Yet, I felt that I had come full circle when, at the end of 1988, all (European) missiles were withdrawn from Europe and also from our country as a result of the INF (intermediate-range nuclear forces) agreement between President Reagan and Soviet leader Gorbachov. I saw this as the vindication of my strategy.

Mobutu

The Cold War was also never far away in the relations between Belgium and its former colonies in Central Africa. During the Second World War the then Belgian Congo had already played a crucial role in the manufacture of nuclear bombs by making its uranium available to the United States. Thereafter, Zaire, as the country was called from the beginning of 1970, became a geo-strategically important ally of the West. However, for a Belgian Prime Minister, Zaire was a very explosive dossier. The steadily worsening situation within the country was no small part of this. Nor was the central figure, Joseph Désiré Mobutu, who developed his own way of dealing with the Belgians.

As Prime Minister, I visited Zaire for the second time in 1988. At that time I regularly met President Mobutu whenever he was in Brussels. Our relations were not altogether smooth. Reports showed increasing evidence of human rights violations and corruption in presidential and government circles. Belgian investors had been frightened off since the nationalisations and the degeneration of the country's infrastructure. During my weekly audiences with King Baudouin, I also noticed that he was increasingly distancing himself from Mobutu. On his last visit to Zaire, to celebrate twenty five years of independence, the King had placed strong emphasis on respecting human rights, to the evident displeasure of the Congolese rulers. This did not go down well with Mobutu, who did not like being told what to do.

On 30 October 1988, I and my Minister of Foreign Affairs, Leo Tindemans, landed at the international airport of Gbadolite. It was my first overwhelming acquaintance with the luxurious city President Mobutu had had built for himself in the north Congolese jungle. The airport was so well equipped that even Concordes could take off and land there.

We were given a warm reception; Mobutu kept calling me his friend but became sullen as soon as the conversation turned to rearranging the debt. Mobutu did not appear to be greatly appreciative of our gesture. "Mr Prime Minister", he said cynically, "we quite clearly owe you 40 billion francs (1 billion Euros) and you are cancelling 1 billion francs (25 million Euros) of this. *Merci beaucoup*". And he went on: "But last year Canada waived our entire debt of 1.5 billion francs (37.5 million Euros) and West Germany promised to wipe out our bilateral debt of 25 billion francs (625 million Euros). And you know, these are countries that never colonised us".

I tried to explain to Mobutu that our gesture was far greater than 1 billion francs, or 25 million Euros. I made it clear that the repayments were ultimately beneficial to his country and that I would also ask Parliament to make the costs of the commercial debts as light as possible. I also referred to our pleas to the Paris Club, the IMF and the World Bank to show more flexibility towards Kinshasa. Unfortunately, Mobutu was not prepared to discuss it: "All that is too complicated for me. I like clear agreements without loopholes".

Old Debts

As had often happened before, Mobutu played the injured party. In his long monologue there was a great deal of bitter criticism of our colonial past: "The press in your country often talks about the violation of human rights. When I entered the service of the Congolese army as a recruit in July 1950, it was not more than a week before I was given eight lashes by your fellow countrymen because I did not have my cap on properly. Violation of human rights? That was the order of the day at the time. It is certainly noticeable that Westerners only discovered the rights of man after they had been chased out of their colonies. And I, 'a violator of human rights', cannot even begin to fill the prisons you built here!"

After my return, Flemish Socialist Members of Parliament declared that the Congolese regime was responsible for the immense problems of corruption, social upheaval and economic recession. They warned that rather than lead to an improvement in the lot of ordinary Africans, the cancelled debts would in fact contribute to the enrichment of the rulers. In addition, it was maintained that they would by now also be skimming off part of our development aid for themselves.

I knew that Mobutu would once again interpret these reproaches as an insult. During our conversations he continually complained about what the Belgians were saying and writing. At the beginning of December, scarcely a month after my return, came the feared response. First, he announced that he was reneging on repaying the debt and was donating all of his Belgian properties to the Zairean state. Every Congolese with property in Belgium must sell it before the end of the year. Further, all Congolese who were studying in Belgium had to leave the country by the end of the academic year.

After that it went from bad to worse. The climax of this "propaganda war" was the announcement that Mobutu was going to set up a large-scale information campaign, to give "a population consisting largely of young people as accurate an idea as possible of the nature and extent of the exploitation and plundering their country had been subjected to for decades". The campaign was to be conducted under the motto "From Leopold II to Baudouin I" and was a response to a Belgian–American television documentary about Mobutu's fortune.

This escalation attracted international interest. At the request of Mobutu, the King of Morocco, Hassan II, offered to act as intermediary. On 6 February

1989, I received an unexpected telephone call. Mobutu was inviting me to his villa in Cannes. I was warmly received because he was very pleased that our French-speaking public TV station had decided not to give the controversial documentary about his reputed fortune a second viewing. We agreed that we would attempt to calm feelings down in order to prepare a round-table conference in a serene atmosphere. We still had a long way to go but eventually, in mid-1989, I did succeed in concluding a global agreement on waiving the debts.

Not a Friend

Our relationship with Zaire slowly got back to normal, although there were alarming reports about the increasing suppression of the opposition. Even though for diplomatic reasons I had called Mobutu "a friend of our country" eight years previously, he was never a friend of mine. At times, he was extremely arrogant and on the telephone he treated me with a certain disdain. He gave me the impression that as a head of state and "Marshal" he felt himself to be superior to a simple Prime Minister. He even professed to be a strategic thinker. But he was authoritarian and militaristic. For him, expressions like freedom of the press, free speech and an opposition were anything but self-evident.

Mobutu denied that he was a politician, although he personally used all the tricks of the trade. Every morning, he listened to Belgian radio and especially to the press review. He also often telephoned Belgian politicians. He was extremely well informed about the vagaries of our domestic politics and took pleasure in exploiting and embellishing differences of opinion, alleged or otherwise. From a secret Blumenthal report by the World Bank it was possible to deduce that Mobutu was buying the goodwill of Belgian organisations and people with gifts and material benefits.

Mobutu attached a great deal of importance to good relations with the Belgian royal family. The already cool relationship with Baudouin hit the rocks completely when, in response to a French-speaking newspaper that had called him a bastard, Mobutu let it be known via his press bureau that "bastards were to be found in particular at the Belgian Court, among the Saxe-Coburgs". This was a colossal and unforgivable blunder, and I let him know this in no uncertain terms. After he had published a hand-written letter from Baudouin, the palace door was shut to him forever. In August 1993, Mobutu, like Saddam Hussein, was struck off the list of heads of state who were invited to the King's funeral. Mobutu confided to a friend that this was the greatest disappointment of his life: "Belgium has deeply offended the Congolese people by this. Baudouin was head of state of the Congo between 1950 and 1960. An absolute disgrace!"

Lubumbashi

A total break came after the Zairean presidential elite troops forced their way into the university campus of Lubumbashi during the night of 11–12 May 1990 and created a veritable bloodbath. Belgium proposed that an international commission should investigate the matter, but Mobutu violently opposed this. He insinuated that we were after his scalp and, to add weight to his words, he suspended all technical cooperation. All the NGOs that were financed by Belgium had to bring their activities to a halt.

Six months later, Mobutu asked for the restoration of non-governmental aid. In the meantime, in the Congo the National Conference, a roundtable with representatives of all political tendencies led by Monseigneur Laurent Monsengwo, was trying to establish democratic structures in Zaire. This was the one and only initiative towards democratisation that Mobutu had undertaken after the end of the Cold War. Because the Conference had a great deal of legitimacy, our answer was that our NGOs could start work again if the National Conference specifically asked them to do so.

Subsequently, however, a struggle for power took place between Mobutu and the Prime Minister appointed by the Conference, Etienne Tshisekedi. Mobutu had the houses of Tshisekedi's supporters blasted by bazookas, and riots broke out in Kinshasa. Civil war seemed inevitable. On 25 October 1991, I called on our 2,000 Belgian nationals to leave the Congo as soon as possible under the protection of the 800 Belgian paratroopers who were operating out of Brazzaville. Our evacuation plan was carried out almost flawlessly and without any major incidents. This was my last decision on the Congo, as the government had to resign after a highly charged dispute over licences for weapons exports. Parliament was dissolved and in the election campaign there was no interest in the escape of our fellow Belgians and the tragic fate of the Congolese people.

Rwanda

A second African region that causes a Belgian Prime Minister great concern is the former mandate regions of Rwanda and Burundi. As government leader, I mainly became involved with Rwanda after the invasion by Paul Kagame's Tutsi rebel army in October 1990.

In the wake of the First World War, when the German empire lost its colonies, Belgium was assigned the guardianship of Rwanda and Burundi by the League of Nations. By contrast with Burundi, where tension between Hutus and Tutsis led in 1972 to terrible massacres with tens of thousands killed, the situation in Rwanda remained peaceful for a long time. Here the previously dominant Tutsi minority had to make way for the Hutu majority. The democratisation of Rwanda, to a great extent supported by a large number of missionaries, seemed to have been successful. Rwanda was internationally regarded as a model country with an original form of development cooperation described as co-administration. Everything seemed to be just fine. And yet, the massive exodus of Tutsis to the neighbouring countries of Congo and Tanzania, and especially the concentration of Tutsi refugees in Uganda, attracted scarcely any attention.

Along with my Minister of Foreign Affairs, Mark Eyskens, I opened the National Museum of Rwanda in Butare at the end of September 1989. The reception I was given there – I was the first government leader to visit the country since independence – was indescribable. The road between the airport and the capital was lined with tens of thousands of Rwandans dancing and singing. The talks with President Juvénal Habyarimana were extremely friendly. I returned to Brussels feeling good. There was nothing to suggest that a year later an event would take place that would continue to fester for four years and end in a horrific genocide.

Bush Warriors

On 2 October 1990, at about mid-day, I consulted as usual the telexes on the government computer, Bistel. The press agency Belga was reporting that a large group of rebels had invaded Rwanda from Uganda. These rebels called themselves the Rwandan Patriotic Front (RPF). They wanted to dislodge the ruling Hutus and form a new government of Tutsis, it subsequently turned out. Several hours after the Belga telex I received "a formal request for assistance" from the Rwandan President. But the Socialist ministers in my government were very reluctant to provide aid to Rwanda.

On Wednesday, 3 October, I was invited by King Baudouin to lunch with Rwandan President Habyarimana, who had travelled on to Brussels from the UNESCO children's summit in New York. The President in no uncertain terms insisted on paratroopers being sent to help his troops drive out the "RPF Tutsi rebels". I said this was impossible. The President's reaction was one of utter surprise at my negative answer. After some discussion, in which King Baudouin played an intermediary role, it was agreed that I would try to convince my government to send paratroopers to protect the numerous Belgian nationals in Rwanda. Their task would be strictly humanitarian.

After the conversation, I convened the inner cabinet, which accepted my proposal to quickly deliver an order of munitions placed by Rwanda with the Belgian firearms company Fabrique Nationale de Herstal (FN). That same evening a C-130 Hercules flew the weapons to Kigali. The next day, the core cabinet decided to send to Kigali 500 paratroopers with military materials as part of a humanitarian mission. Their task was to safeguard Kigali Airport and the approach roads and to gather, protect and escort the Belgians in Rwanda if necessary, with a view to possible evacuation. All this had to be done in close cooperation with the French forces, who had also flown paratroopers to Kigali.

On Sunday, 8 October, I received President Yoweri Museveni of Uganda. He was also chair of the Organisation of African Unity (OAU). I confronted him with the reports that the rebels were coming from his country and, moreover, had served in the Ugandan army. The President told me that in the last thirty years hundreds of thousands of Rwandans had fled to his country and also to Zaire and Tanzania. At the end of our conversation he said with a great deal of emphasis in his voice: "I wanted to see you to advise you that these are experienced soldiers, some have been fighting for as much as twenty years, they are "bush warriors". Uganda will not let them back in because we are fed up with their problems. There is no military solution to the difficulties in Rwanda. The problems can only be solved politically. We must convince Habyarimana of this. A solution is urgently needed because the problems in Rwanda may escalate. Personally, I am disappointed at the lack of progress because I have been very active myself".

The conversation with Museveni made me decide that it was necessary to take the diplomatic route, and not the military one, to keep the Rwandan conflict under control. Meanwhile, my Socialist government partner and the opposition inundated me with requests to withdraw our paratroopers as quickly as possible. King Baudouin warned of the pernicious consequences of a premature withdrawal of our troops from a friendly country where, thanks to our aid workers and our historical bonds, we had taken upon ourselves a great responsibility for development.

The Peace Mission to the Great Lakes

We could not reach agreement in the government. Two of my five Vice-Premiers wanted to bring back not only our paratroopers but also our development workers as quickly as possible, as had happened a few months previously in the Congo. They thought we should bid farewell to our African past. But I agreed with the King that the sudden withdrawal of all our workers would cause the complete collapse of Rwandan society.

On Saturday evening, 13 October, my Vice-Premiers and I came to the decision that only an international diplomatic initiative could defuse the Rwandan issue. We decided to send a delegation under my leadership to the region to enter into negotiations with all the parties involved. We wanted to achieve a ceasefire followed by a monitored truce. There had to be political dialogue between the government and the rebels and an international conference had to be held.

The very next day the Belgian peace mission left for the Kenyan capital Nairobi, where we first had talks with President Habyarimana and his Minister of Foreign Affairs, Casimir Bizimungu. I told the press that as long as the negotiations continued, the paratroopers would remain in place and the evacuation of the Belgians would be deferred. The following day, we conducted a kind of shuttle diplomacy with the presidents of Tanzania, Kenya, Uganda and Burundi. All showed themselves willing to subscribe to our main principles and objectives.

The agreement that was signed by Rwanda and Uganda on Wednesday, 17 October, in Mwanza at Lake Victoria in Tanzania showed promise but was not very conclusive. According to the wording, a dialogue had to be instigated with the internal and external opposition. Efforts had to be made to reach an armistice, which would have to be safeguarded by foreign troops. Finally, a regional conference of the Great Lakes, including the countries of Rwanda, Burundi, Zaire, Uganda, Kenya and Tanzania, would have to sort out the acute refugee problem.

A controversy later arose over the composition of the foreign military force that was to oversee the ceasefire. In my talks with President Habyarimana, the idea had developed that a European peacekeeping force might play a role in Rwanda. This peacekeeping force would initially consist of Belgian and French soldiers under the umbrella of the twelve European Community Member States. We had discussed this idea, expressly stating that all our European partners had to consent to it. President Habyarimana had responded enthusiastically. In a conversation with French President Mitterrand, Habyarimana underlined that I was the father of the idea of a European peacekeeping mission. A few days later Minister Eyskens sounded out his European colleagues on this line of thought, but they politely answered that the European treaties did not allow for such a mission. It would take fifteen more years and hundreds of thousands of victims before European troops would be allowed to be deployed for peacekeeping missions in the region of the Great Lakes. This experience also explains my aversion to formalistic and bureaucratic diplomacy.

To make this truce permanent, I set off once more with my Ministers of Foreign Affairs and Defence for Nairobi, where we negotiated intensively for two days with representatives of the OAU, who were able to contact the invading troops of the RPF. Just before our return to Brussels, on Wednesday, 25 October, we were able to announce that the warring parties had negotiated a truce. It came into force at ten O'clock. It was worrying, though, that the truce agreement was immediately violated. It also worried me that we had seen or heard nothing of President Habyarimana. He had gone with his colleague from Burundi to Mobutu's residence in Gbadolite for parallel negotiations. Mobutu had greatly resented the fact that he was the only head of state from the region who had not been involved in the agreement of Mwanza. But after our criticism of the bloodbath at Lubumbashi, Mobutu himself had broken off cooperation with Belgium.

Back in Belgium, we decided to bring back the paratroopers on 1 November. The agreements of Mwanza were unfortunately never carried out. The Rwandan government broke the ceasefire and started a great offensive, probably with the aid of Zairean weapons and troops. The rebels withdrew into the neighbouring countries or went underground in Rwanda itself. I continued to observe this development with a feeling of disappointment and was forced to admit that African negotiators did not honour their word.

Mobutu thereupon took over the diplomatic initiative. His efforts led to a new truce in March 1991, the agreements of Kinshasa in 1992 and, in the summer of 1993, to the agreements made in the Tanzanian town of Arusha. Here, the extremely complicated and detailed division of authority and posts in the Rwandan bureaucracy was agreed on. After my resignation as Prime Minister in March 1992, I no longer played an active role in this matter.

A Tragic President

As a member of the European Parliament and EPP President, I was in Kigali again in February 1994. At that time there was still hope that the Arusha agreements would be carried out by the Rwandan transition government. Just in case, the United Nations had sent a contingent of blue helmets there to monitor the peacekeeping. In addition to 400 Belgian paratroopers, the UN contingent was made up of Bengalis, Ghanaians and Senegalese.

I was in contact there with Habyarimana, the then Prime Minister, Agathe Uwilingivmana, and also with Prime Minister–designate Faustin Twagiramungu. I attempted to establish clarity among the chairs of the fiercely rival political parties and of the highest court of law. Many of the people I spoke to were murdered a few months later. I was clearly struck by two things: the polarisation of minds and the isolation of the President. I noticed that Habyarimana had a very poor relationship with his Prime Minister, Uwilingivmana. She was a progressive Hutu and her political influence was growing, whereas that of the President was waning.

At the request of Habyarimana, I also went to see the Tutsi rebel leader Kagame at his headquarters in Mulindi, an ugly hill village in the north of Rwanda. It was an adventurous journey, not without danger, accompanied by Belgian blue helmets and RPF soldiers. We went through enemy territory and passed dozens of roadblocks manned by heavily armed militia. An estimated 300,000 refugees were living in the camps along the road. I sensed that the situation was very explosive and fragile.

I had a message with me for Kagame, in which Habyarimana confirmed that he was indeed ready to faithfully carry out the Arusha agreements. I additionally needed to sound out Kagame as to whether he did genuinely want elections for local councils, parliament and the president. Habyarimana doubted this because he believed the RPF would definitely lose. Furthermore, the Rwandan President wanted the warring parties to sign a "code of good conduct" to prevent acts of revenge and massacres and to have everything run by means of democratic decision-making. In fact, what it amounted to was that no persecutions would be instituted for violent actions that happened prior to the Arusha agreements.

Kagame responded very condescendingly to Habyarimana's message. He thought the code useless. The conversation became very difficult, almost hostile, and to add to this, was conducted in very poor English. Afterwards, I also visited various refugee camps in the south of Rwanda. In Kigali I was interviewed by journalists from Radio Mille Collines. The questions were fired at me quite aggressively, and I was held responsible for the "Belgian betrayal of 1990".

When I reported this to the President, he responded dejectedly. It seemed to him that all was lost. To me, there was an air of tragedy about him. He feared that as President he would no longer have any real power following the mechanisms set out in the Arusha agreements, which specified strict majorities in Parliament and gave the government a deciding vote. During a last conversation with him alone it again became painfully clear to me that he was increasingly isolated and that he was no longer pulling all the strings. But I always believed in his honesty. The assumption that at that moment he was preparing wholesale slaughter against his opponents was inconceivable to me.

The rest of the story is well known. At the beginning of April the aircraft with the Rwandan and Burundian Presidents on board was shot down. A few days later Prime Minister Uwilingivmana and her children were murdered, as were the ten Belgian blue helmets that were responsible for her safety. Subsequently, the Belgian and most of the other blue helmets were withdrawn, and our nationals in Rwanda were evacuated under military escort. In the meantime, the Tutsis and the moderate Hutus were subjected to an unprecedented bloodbath. Kagame's RPF subsequently captured the whole of Rwanda and carried out massacres. Many Hutus fled to neighbouring countries where they have since been involved in bloody conflicts in the area of the Great Lakes for more than ten years.

Pilgrimage Through Africa

I was too heavily involved in this drama to be able to make a definitive final judgement. But I am convinced that much shedding of blood and ethnic violence could have been avoided if the West had taken stronger action. King Baudouin spoke prophetically when on 10 October 1990 he warned that a premature departure of Belgian paratroopers would greatly increase the risk of inter-ethnic massacres in Rwanda and perhaps also in Burundi. I think that, if they are honest, many political leaders must be suffering from a guilty conscience over this event.

My concern about Africa has only become stronger since the end of my premiership. Prompted by Médecins sans Frontières, I went to Somalia in August 1992. When I first landed there, in the company of a medical team, we were confronted by refugee camps such as that at Mandera on the borders of Kenya, Somalia and Ethiopia. The UN High Commission for Refugees had an impossible task there. I saw starving children dying in the arms of their mothers. This had such a powerful effect on me that I could not hold back the tears. Since then I have carried those children in my heart.

We generally stayed in quite dilapidated, sometimes very primitive houses where Médecins sans Frontières had set up its quarters. We sometimes ate food from military rations, in packs which we tore open like astronauts. In Kismayo we slept on the ground. On Christmas Eve 1992, I was sharing a room with the Belgian ambassador who had come with me from Nairobi. We had no sanitation, but we did have a mosquito net. It was unbearably hot and all night long we could hear the gunshots of the Somali rebels. I remember with great disdain the shortage of water. On 31 July 1993, I was again on my way to Somalia via Kenya. In Nairobi early in the morning I heard of the death of King Baudouin. I returned for his funeral and the taking of the oath by the new Belgian King Albert II, after which I left for Kismayo again. The Belgian military had successfully brought about a regional peace agreement there.

I then took a small plane from northern Kenya, where the United Nations and Médecins sans Frontières were running a few large refugee camps, and flew to southern Sudan, because it was also suffering from famine as a result of the civil war. Fortunately, we were not shot at during our flight. I thought we were flying to a desert region but it turned out to be a green and fertile country on the upper Nile. A Médecins sans Frontières team was trying to set up a hospital there in the most rudimentary circumstances.

After Sudan, I flew to New York and met with the UN Secretary-General, Boutros Boutros-Ghali, to get him finally to put this terrible civil war on the world's political agenda. He answered me in all honesty: "You must mobilise your government leaders and public opinion in Europe, otherwise there is little I can do". He also said that the government in Khartoum "is a government of fools".

In 1993, once again at Christmas, I found myself in the almost intolerable horror of the Angolan civil war, in Malanje, a besieged town that was forced to live like Berlin during the airlift. The children who used to search the airfield for a few lost grains of corn were beaten by the police and chased away. The peace agreement in 1992 between the two Angolan rivals, Unita of Savimbi and the Movimento Popular de Libertação de Angola of Dos Santos, to make possible general elections, had failed due to poor judgement: unlike Mozambique, neither the warring armies nor the police forces were dissolved beforehand and the country had not been cleared of mines. I have nowhere in the world seen so many young people with amputated limbs as in the Angolan capital, Luanda.

My pilgrimage to Angola, Burundi, Ethiopia, Kenya, Namibia, Rwanda, Sudan, Somalia and South Africa with the noble teams of Médecins sans Frontières, who were trying to do something about the greatest human need in the area, lasted more than two years. I also went to the African continent several times for the purpose of promoting democratisation. On Whitsun 1993, I was in Malawi, where a campaign was being conducted for a referendum on a multi-party system. I went to support the proponents, who subsequently won. I then travelled to Mozambique as a foreign observer during the elections. In 1994, I was an observer at the first free elections in South Africa.

There is a long way to go for the European Union. At a symposium in Brussels on 18 March 1994 on the theme, "Africa is dying, do we care?", I spoke about my experiences on the continent and concluded that "Europe must not itself contribute to the deterioration of competition between the developing countries. On the contrary, Europe must contribute to a new economic world order through greater openness and through respect for certain standards of fairness. In short, Africa may rightly expect Europe to clearly and plainly promote deeds of justice and solidarity".

It is in all respects a sign of hope that international celebrities are using rock festivals in an attempt to alert the world and have its leaders put poverty in Africa on the international agenda. Hopefully, waiving the debts of the poorest countries was not a one-time event, intended only to absolve the rich West of a guilty conscience.

Chapter V

My Road Towards Maastricht

One of the principal ongoing tasks of a Belgian Prime Minister, besides trans-Atlantic relations and cooperation with the former colonies in Africa, is of course Europe. I was a member of the European Council of Heads of State and Government for more than twelve years. Initially, due to my age and limited experience, I was very much the newcomer, but I gradually became one of the more familiar figures, thanks to the central role of Belgium in the process of European integration and my long record of service.

The Giscard–Schmidt Tandem

When I became Prime Minister in April 1979, France held the Presidency of the Community and Valéry Giscard d'Estaing acted as host. Margaret Thatcher, the newly elected Prime Minister of the United Kingdom, and I were the two newcomers to the European Council, which at that time numbered only nine members. By a quirk of the alphabet we found ourselves sitting next to one another. Only a month into my premiership, I signed, together with my European colleagues in Athens, Greece's accession as the tenth Member State.

Giscard's summit, on 21–22 June, began with a kind of procession, wandering through the streets of Strasbourg from one conference site to the next. The exchange of views still took place very informally. There was far less paper in circulation than there is today. Away with this *Dreck,* or crap, as Helmut Schmidt referred to the piles of paper. Translation booths already existed, but he generally spoke English. Now everyone speaks his or her own language. Moreover, interventions were kept short, at Giscard's firm insistence. Opinions on international politics were often exchanged largely over dinner. As a newcomer I adopted a rather discreet attitude.

Giscard and Schmidt dominated the gathering, but because of the small number of participants real dialogue remained possible. However, these conversations did not lead to concrete results, because no conclusions were drawn, let alone applied. In 1979 the energy problem and the battle against inflation were already high on the agenda. But at that time any kind of mutual approach was still a long way off. The Community was not yet authorised to deal even with energy policy.

The European Council may have been informal, but it has always been crucial for top appointments, in particular that of President of the Commission. This first became clear to me at the Venice Summit on 12–13 June 1980. At that summit, Francesco Cossiga, the President of the Council, Prime Minister and later President of Italy, organised all the individual private consultations with each government leader to find a successor to Roy Jenkins, the British President of the Commission who was stepping down. My suggestion to Cossiga and also to Giscard was Leo Tindemans, known as "Mister Europe". But the French President could not be persuaded. In fact, neither could Tindemans, even though he was the most prestigious candidate in the Benelux and could count on the support of Helmut Schmidt. It was a great pity he was unwilling to stand. Perhaps he was hoping to become Prime Minister of Belgium again. The private consultations therefore failed to come up with a result. At a subsequent European summit the Prime Minister of Luxembourg, Gaston Thorn, was appointed President of the Commission.

First Time in Maastricht

The Dutch town of Maastricht, which had still briefly belonged to Belgium after our country became independent, was the site of two historic European Councils. The summit that produced the Treaty of Maastricht in December 1991 is well enough known. But the first summit, held ten years earlier on 23–24 March 1981, was also crucial for me. As I described in Chap. III, "In Charge of a Restless State", our country was regarded as the sick man of Europe at that time. We had a thoroughly bad reputation. European institutions sharply criticised the fact that we had failed to make any radical reforms. As an advocate of a drastic change in direction, I sought the support of my European colleagues in the battle I was waging on the home front. The question of the site of the European Parliament was also raised. The fight over the permanent site for the Parliament was to be hard and long. In the run-up to the French presidential elections, Giscard attempted to force the issue in favour of Strasbourg. At the time, the Parliament was housed in both Brussels and Strasbourg, and MEPs commuted between the two cities. Fortunately, the extraordinary sessions in Luxembourg had been abolished in 1981.

There are historical, political and moral arguments for maintaining Strasbourg as the home of European institutions. But on grounds of democracy and efficiency I have always been and still am in favour of Brussels as the exclusive site for Parliament, Commission and Council, primarily in order to enable Parliament to exercise efficient control. That has always been the will of the majority of MEPs. It is beyond question that the "Community institutions" are also a particular trump card for Brussels and Belgium. But there are other European organisations based in Strasbourg, such as the Council of Europe and the European Court of Human Rights, which can compensate for the loss of the Parliament.

However, it was not that simple. Successive French presidents have set themselves up as defenders of Strasbourg, supported in this by the Germans. Even in 1981 I would not budge, for which Giscard did not thank me. The compromise, which was to take a decade, provided for plenary sessions to be held in Strasbourg and political groups and committee meetings in Brussels. But this status quo, established at Maastricht, was actually very inefficient, logistically and financially.

My First Presidency (1982)

In the first six months of 1982 I was President of the European Council. At the Brussels Summit of 29–30 March I was able to tell my colleagues about the drastic measures for recovery my new government had taken. Helmut Schmidt was enthusiastic about the change. He addressed me by my first name for the first time. The others explained what they were doing in their own country to combat the spectre of inflation and unemployment. Mrs Thatcher found this very interesting and congratulated me. However, in my view it lacked synthesis and a common approach. Not just the same objectives, but also mutual resources were necessary. In Mrs Thatcher's view, however, this was going too far. She could not understand why I kept hammering away at the importance of "Community institutions", especially the Commission. She said, "Your government and mine are good governments. We pursue good policies. That is what is important. Not the Community institutions".

As chair of the European Council, I was invited by President Mitterrand to the Western Economic Summit, the forerunner of the G-7 and G-8, held at Versailles on 4–6 June 1982. In the classic "family photo" I stood at one end of the row and Commission President Gaston Thorn at the other. However, in the international newspapers we were both cut off! France's economic policy, directed by Prime Minister Pierre Mauroy, who headed a government of Socialists and Communists, was a sore point at the meeting. There I met his Minister of Economics and Finance, Jacques Delors, for the first time. A careful eye was being kept on the economic policy of the French and was scrutinised by Washington, London and Bonn. After the summit, Paris quickly backed down and finally chose an economic model more oriented towards the social market.

At the second European Council meeting under Belgian Presidency, on 28–29 June 1982, enlargement to include Spain and Portugal was up for discussion.

Negotiations promised to be difficult and long transition periods had already been anticipated. The poor functioning of the Community, and doubts about the Community's capacity for integration after the enlargements of 1973 and 1981, begged the question: Had the entry of the United Kingdom, Ireland, Denmark and Greece reinforced integration? It was difficult to answer this question definitively. Even then, deepening versus widening was already being debated.

Towards a Re-Launch

The first seeds of revival were beginning to sprout in the mid-1980s. A significant obstacle was negotiated during the Fontainebleau Summit on 25–26 June 1984. After years of obstruction, François Mitterrand, in consultation with Helmut Kohl, reached an agreement with Margaret Thatcher on the "rebate": the partial reimbursement of the "membership fee" (the relatively high net contribution of the United Kingdom to the Community). This brought an end to the controversy provoked by Mrs Thatcher five years earlier when, at the Strasbourg Summit, she had cried out, "I want my money back!"

The summit also had to appoint a successor to Commission President Gaston Thorn. Traditionally, this matter is handled informally during and after dinner. The French proposed their Minister of Foreign Affairs, Claude Cheysson. But he had stood as a *candidat socialiste* and had in France advocated cooperation with the Communists. For this reason, he had no chance to be appointed since the decision required consensus. After the failure of Cheysson, I attempted to have my fellow countryman Steve Davignon chosen, as he was generally extolled as a brilliant Vice-President of the Commission. Helmut Schmidt always spoke with great praise for "the Viscount". Yet this still did not work, because he had lobbied too openly for himself. Jacques Delors was then appointed at the beginning of July. This was pre-eminently a Franco-German nomination, but the French candidate was also acceptable to Margaret Thatcher, due to his market orientation. She later came to regret it.

At the beginning of 1985, Jacques Delors took office as the new President and succeeded in getting European integration back on track. He was resolutely determined to create an internal market. On 14 June the late Lord Arthur Cockfield, a British Conservative and Commissioner for Internal Market Tax Law and Customs, proposed all the measures necessary in his White Paper, "On Completing the Single Market". Delors put the document before the European Council on 28–29 June in Milan.

There was heated discussion about holding an Intergovernmental Conference (IGC), which is necessary to negotiate a new treaty. "You cannot decide that", shouted Thatcher, "because I oppose it". The Italian Minister of Foreign Affairs, Giulio Andreotti, who was jointly conducting the meeting with Prime Minister Bettino Craxi, immediately picked up the Treaty of Rome and quoted Article 236, from which it appeared that a decision could be made on the holding of an

IGC by a simple majority. Thatcher was dumbfounded. Everyone voted in favour; she was the only one who voted against and the Danes abstained. But during the summer, Mrs Thatcher changed her attitude and began to join in the preparations for the new treaty. She could now see herself that a unified European market was the natural progression of her own economic convictions, and that turning it into reality would definitely be to the advantage of the British economy. At the Luxembourg Summit on 2–4 December, agreement was reached on the Single European Act. On 1 July 1987, it came into force. Because of the efficient preparations by the Commission and because the government leaders were brave enough to take the plunge, a re-launch of Europe was accomplished in the following six months. "Eurosclerosis", which had lasted for years, had come to an end.

Many years later, I once again broached the subject of the Single European Act during a debate with Margaret Thatcher. She, together with George H. W. Bush and Mikhail Gorbachov, had been invited to a celebration for a leading Belgian businessman. Gorbachov had cancelled at the last minute due to his wife's illness. A few hours before the celebration I received a phone call and was immediately connected to President Bush, who asked me to replace Gorbachov in the debate. It was conducted by the well-known BBC journalist David Frost, who asked us who the greatest politician of the twentieth century was. Mrs Thatcher answered Winston Churchill, Bush said Ronald Reagan and I chose Jean Monnet. This gave rise to an animated discussion with Mrs Thatcher, during which she brought up the Eurosceptic arguments that have since become her trademark. I reminded her that she had nonetheless agreed to the European Single Market in the summer of 1985. "Yes", she said, "and it was my greatest political mistake!"

My Second Presidency (1987)

Belgium once again held the Presidency of the Community in the first six months of 1987. A visit to the United States had also been scheduled for me. On 28 May I met President Reagan in the Oval Office. I handed him a box containing one gold coin worth 50 ECU (European Currency Unit) and one silver coin worth 5 ECU, a gift that he very much appreciated, although I do not think he was aware of our ambition to create a single European currency. The conversation did not last very long, but was most congenial. Reagan broached the subject of the American economy. He laid the blame for the excessive budget deficit on the Congress. He also hammered away at the necessity to lower taxes – "low tax rates generate high income" – and cut back state interference in the economy, "deregulation". In short, the classic tenets of "Reaganomics", which were not very popular in continental Europe.

Reagan was troubled at the time by the Iran-Contra affair; Mitterrand was in the midst of his first *cohabitation* with Jacques Chirac (Mitterrand, a Socialist President, had appointed a Conservative Prime Minister) and was far from certain of being re-elected President in 1988; Kohl's CDU had suffered heavy losses in some state elections; the Italian government, as ever, was hanging by a thread; Mrs Thatcher was facing crucial parliamentary elections and the position of Yasuhiro Nakasone, the Japanese Prime Minister, had been weakened by persistent economic problems. Everyone was very much in need of a successful Western Economic Summit in Venice. I stuck to my European role. Tasks were divided between the Council and the Commission. Some dossiers were dealt with by me and others by the President of the Commission. For Delors, participation in this summit was a matter of prestige and proof of the global importance of the European internal market. He believed a central role should be reserved for him and therefore threatened not to leave for Venice unless he were involved in the financial and monetary discussion of the seven, although it is doubtful he would actually have carried out this threat.

The meetings, dinners and lunches took place on three parallel circuits, each with their own formal and informal agendas: the first involving the heads of state and government leaders, the second the ministers of Foreign Affairs and the third the ministers of Finance. Most of the activities had been prepared well in advance by the advisors or "sherpas" of the government leaders and also by Delors' chief of staff, Pascal Lamy, later European Commissioner and currently Director-General of the World Trade Organisation. The formal gatherings of the Western Economic Summit consisted mainly of final discussions based on the notes drawn up by the sherpas. Strictly speaking, no negotiations took place and no decisions were made. Yet a summit of this kind was and still is important. Thought is given at the highest level to the most important economic problems, and attempts are made to achieve at least some coordination of policy.

Summit meetings of this kind are just the tip of the iceberg. During the six months of the European presidency initiatives were taken, agreements forged and successes recorded in a large number of areas. I myself, for instance, pressed for a European social dialogue. Up until then the social partners had been absent at a European level. But with the prospect of the completion of the internal market in view, it was time for the social component to catch up with the economic. That was why on 7 May, Delors and I invited representatives of the European employers and employees to an official dialogue. It was a first in the history of European integration.

On a Pilgrimage for Europe

As always, the highlight of the presidency had been reserved for the closing European Council. In the course of a few days I visited all the capitals of the then Community. In *Die Welt* of 22 June 1987 I described my tour as a *Pilgerfahrt für Europa*.

This time, a common political and economic project did come up for discussion at the European Council of 29–30 June. In a document entitled "Making a Success of the Single Act: A New Frontier for Europe", Delors proposed a drastic reform of the budget, with a larger proportion going to projects that would enhance cohesion, so that credits for the Common Agricultural Policy would drop in relative terms. This met with heavy opposition from Margaret Thatcher. In a very caustic debate, Mitterrand and Kohl insisted on an agreement. However, when the time came to vote, I reversed the voting order so that Margaret Thatcher came last. Everyone voted in favour apart from her. She was totally isolated, but because unanimity was required, no agreement could be reached. Due to the action of the British there was also no agreement on the Belgian proposal to draw up a core of basic social rights, later called the Social Charter. Because I could see the storm gathering and wanted to prevent a formal rejection, I stopped Foreign Secretary Geoffrey Howe in his tracks and called out: "Bring in the sandwiches!" The summit had nonetheless failed.

The Danish Presidency in the second half of 1987 could reach no consensus either. Kohl, who was presiding in the first half of 1988, managed to succeed in this at the Brussels Summit of 11–12 February. Agreement on the first multi-year budget, known as the Delors package, gave a financial boost to integration and contributed to the euphoria of "Europe 1992", as the completion of the internal market was called.

Ambitions for a Top Job

Owing to the growing prestige of Europe, membership in the Commission was much sought after. In Belgian tradition, decisions were made on this during governmental negotiations: Willy De Clercq, for example, became European Commissioner in 1984 during my time in office. In May 1988, Jean-Luc Dehaene was negotiating with the Socialists and I was asked whether I hoped to join the Commission. Karel Van Miert, then chair of the Flemish Socialists, had twice asked me whether I had ambitions in this direction. "I do not want to replace Willy De Clercq," I said. "It is a matter of loyalty. I would not do it". My political life would have turned out differently if I had said yes. Ultimately, Van Miert was put forward as candidate after the formation of the government.

Speculations regarding my ambitions for a top European appointment cropped up regularly in Parliament. There, in one of the most savage centres of gossip in the world, I overheard a rumour that I was to succeed Delors. I immediately denied that I was after his job. Fortunately, the matter was quickly forgotten. In 1993, the mandate of the Delors presidency was extended by two years until 1 January 1995, so that his term of office coincided with that of the Parliament. Even though there was already an informal consensus that it would next be the turn of a Christian Democrat from a small country, neither Lubbers nor I became President of the Commission. Lubbers was not supported by Kohl, and I did not stand a chance because I was no longer a member of the European Council. If I had remained Prime Minister, it might have been a different story. Helmut Kohl had confided to me, when I resigned in 1992, "You really ought to be President of the Commission. You are my candidate". At the beginning of 1994, at a CDU congress in Hamburg, he told me that he had been sorry to hear "dass du aus dem Rennen bist" (that you are out of the race). Candidates only stand a chance if they are put forward by their own government. Officially, and to conform with coalition agreements, Van Miert remained the official Belgian candidate, but in practice Jean-Luc Dehaene was already fully in the running for the presidency of the Commission.

The Long Way to the Euro

The end of Eurosclerosis was also felt in the EPP, where there was a particularly strong consensus on the future of Europe. At the EPP summit held at the chancellery in Bonn on 30 May 1988, lengthy discussions had already taken place on the link between the internal market and monetary union. Kohl was a great proponent of the European Monetary Union (EMU) and believed that European integration should take on an irreversible character.

Our ambitious and partly prophetic concluding statement went as follows:

> The next step that must be taken towards a European Union by 1992 is the completion of the great, free internal market and its social framework. Founding the European Union remains the task and the goal of Community policy. To achieve this, decisions and measures are necessary that go beyond the reforms ratified by the Single European Act, notably extending the legislative and controlling powers of the Parliament and also the executive powers and initiation rights of the Commission; creating the conditions for founding a European Central Bank to keep autonomous watch over the value and stability of a European currency; further development of the European Political Cooperation into a procedure of Community foreign policy; summarising the current regulations and procedures provided for, within or outside the treaties, and further developing them into a democratic constitution.

We were already proposing a European constitution in 1988!

The European Trio ...

The re-launch of Europe was due to the collaboration among the team of Mitterrand, Kohl and Delors. The other government leaders were not involved in their decision-making, though I was certainly one of those who always supported the *troika* and Delors' leadership, together with Santer of Luxembourg, Andreotti of Italy and Gonzalez of Spain. The atmosphere at the European Council was unique, because so many members were fundamentally in agreement with one another about finalising the EU; moreover, they got on extremely well with one another on a personal level.

I had a very good relationship with Mitterrand, for example, in spite of the ups and downs surrounding the question of the site of the European Parliament. He invited me to the Elysée Palace on various occasions and in 1982 he invited me to Fort Brégançon, a presidential holiday residence. I was on altogether easier terms with Mitterrand than with Giscard, who, each time we met, started the conversation superciliously with the question: "Comment va la Belgique?" Yet there was always a distance between Mitterrand and myself because of his status as head of state, and I always called him *Monsieur le Président*. Only Helmut Kohl and Margaret Thatcher addressed him by his first name.

I always got on extremely well with Helmut Kohl. Our convictions on Europe are identical. I had already known him for years. As the newly minted chair of the Flemish Christian Democrats, I attended the CDU congress on 12 June 1973, where he was elected chair. At the time he was still Minister-President of Rhineland-Palatinate. I went to all the CDU congresses and, as President of the EPP, have continued this tradition right up to the present day. We were members of the European Council together for almost ten years. We always worked together closely and with mutual trust, not only on the Council but also within the EPP. Once you gain his trust, his support is unconditional. But once you forfeit that trust, he drops you like a brick, and Kohl's political "weight" is sufficient to ensure that you are then politically dead.

Delors was in essence a Christian Democrat, even though he belonged to the socialist family. He is a practising Catholic and knows the classic tenets of Christian Democracy – personalism, federalism, subsidiarity – better than many politicians who call themselves Christian Democrats. Moreover, he is exceptionally pro-European, something that cannot be said of all Socialists or French people. In spite of the rumour about my interest in the presidency of the Commission, our relationship has always been excellent. It is remarkable that, to my knowledge, he has never expressed an opinion on the question of the site of the European Parliament. We always, almost instinctively, took the same line. Because of the proximity of the Commission in Brussels, it was exceptionally easy to keep in contact. He fulfilled his role at the European Council with vigour. It was Delors who mapped out the lines along which discussions were to take place. At the same time, he would always go out of his way to thank and congratulate government leaders who had stepped into the breach for him, as I repeatedly found out.

... Versus Mrs Thatcher

I never avoided confrontation with Margaret Thatcher. During a lunch at the European Council in Dublin in June 1990, I spent an hour discussing the future of Europe with her. It was a continuous back-and-forth battle and none of the other government leaders attempted to join in the conversation. They were probably quite happy to let me explain the matter on their behalf.

From the mid-1980s the Franco-German axis, or more precisely the triple team of Mitterrand, Kohl and Delors, worked fine as long as Thatcher did not put a spike in the wheel. If the French and the Germans made a proposal that had been carefully thought out beforehand, it had every chance of success. Preparation was crucial and the British needed to be convinced in advance. This was always a tricky business. Thatcher never hesitated to stick her neck out if talks were not going the way she wanted, even though she was often alone in this and it contributed to her isolation. I doubted whether this was an effective way of conducting European politics, but you had to at least admit that you knew where you stood with her. There were no spin doctors at that time. Her forthright and headstrong behaviour brought her a great deal of sympathy, at least on the home front, but made her many enemies elsewhere.

Her address at the opening of the academic year of the European College in Bruges on 20 September 1988 is legendary. The speech caused quite a stir in European circles. Just as the European project began to take off again, Mrs Thatcher lashed out fiercely against the growing "busybodying" of the Commission, which she depicted as a nitpicking regime, "a nightmare" of a super-state of Europe, which she casually compared to the Soviet Union. She also opposed the founding of a European Central Bank and the abolition of internal borders. It was in fact a rearguard action, since most of her views went against what she had previously approved. Moreover, she refused to acknowledge the logic and dynamics of European integration. An economic union cannot succeed without a monetary and political union. In the evening, following her address, there was a dinner at the British embassy in Brussels. In polite terms I used the opportunity to launch into a passionate discussion. Mrs Thatcher, however, arrogantly expressed her opinion on the political system in Belgium. "How can you keep a country together in that way?" she asked rhetorically. "What on earth are all those districts and communities? Are they *provinces*?" It was hopeless.

I was determined not to leave it at that. Two weeks after the address at Bruges, on 28 September, I arranged an international, well-attended press conference in Brussels in order to explain my federalist view of the future of Europe in response to Thatcher's sharp Euroscepticism. I was not aiming at a personal polemic, but I was pressing for a debate on the fundamental issues. I was the only government leader at the time who took up the gauntlet against the Iron Lady. I, like Thatcher, had more than ten years' experience on the European Council and my European commitment, conviction and views were of long standing.

Some British people, especially a number of British Conservatives, feel closer to the United States than to the European Union. That is the crux of the problem. Margaret Thatcher was also convinced or became convinced that this was so, especially after her mandate as Prime Minister came to an end. She realised far too late that the internal market contained the seeds of economic and monetary union, including a single currency, and ultimately also of political union. She could not come to terms with this. For her the transfer of sovereignty or, more precisely, the sharing of sovereignty, was and still is undesirable, impossible and unthinkable. She was against the expansion of the "Community institutions" and found the European social model horrendous. She fostered a fundamental revulsion for what we had achieved on a social level in continental Europe. It was completely beyond her comprehension and she showed utter contempt for it. The same was true of the organisation of the political system. She lived by the British majoritarian system. Other systems were by definition non-democratic and inefficient. When once I made it clear to her that when making crucial decisions I always had to take coalition partners into account, she answered laconically, "You have got the wrong system!"

Socially Margaret Thatcher was always friendly, however. I liked her and admired her perseverance and enormous capacity for work. When at the end of 1990 her departure from the political stage seemed inevitable, I still encouraged her personally not to lose heart when we were sitting next to one another at the gala dinner in the Hall of Mirrors at Versailles during the summit on the European Charter. But we never arrived at a *rapprochement* of ideas.

The Fall of the Wall

The completion of the internal market and the prospect of a single currency took on a new urgency in the second half of 1989, owing to the upheavals in Central and Eastern Europe. Initially, the two were separate issues. On 28 September 1989, I had an audience with Council President François Mitterrand and his Minister of European Affairs, Elisabeth Guigou, at Stuyvenberg Castle. Mitterrand informed me of his bilateral contacts with Mrs Thatcher. She had warned him of the Socialism that might creep in via the Community "through the back door". He also asked me to work on Lubbers who, in his opinion, relied too heavily on Thatcher. Now that the end of the Cold War was approaching there could still be no question of enlargement. Mitterrand summarised this as follows: "We have many problems to resolve and we need at least fifteen years to tackle them. I will not promote new membership applications because Europe will then break down". In retrospect, his timing seems prophetic. He was expressing the general conviction that we should not let ourselves be thrown into confusion by what was going on outside the Community, for fear that this would bring integration to a standstill.

When the Berlin Wall fell on 9 November 1989, the situation changed completely. It was clear to everyone that the *Wende* ("change") was irreversible. Even the French Presidency came round and reacted very promptly and tactically by convening an informal summit on 18 November at the Elysée. Belgium immediately supported the French initiative. With this first summit meeting since the fall of the Wall, Mitterrand attempted to take the wind out of the sails of Thatcher; he wanted to focus the discussion at the European Council on Central Europe. For Mitterrand it was essential that the Community speak with one voice so as to pre-empt the East–West Summit of Bush and Gorbachov in Malta on 3–4 December.

The informal dinner – which Lubbers ironically called the gastronomic summit – did not formulate any decisions, but it did produce some politically important statements. For instance, explicit support was declared for the reformers in Central and Eastern Europe, with Mitterrand's proposal to establish a European Development Bank. Thatcher opposed this until it was suggested that the bank should be based in London. We also clearly stated that these new developments should not be allowed to slow down European integration. Kohl was fully committed. This was of vital importance, since the French were afraid that a new, enlarged Germany would turn its back on Europe. The inviolability of the existing borders was formally confirmed, as were the military alliances, both NATO and the Warsaw Pact. To my knowledge, German unification or *Wiedervereinigung* was still not mentioned in Kohl's intervention at the time.

No one really knew where we were going. We were mainly feeling our way during the talks. The fall of the Berlin Wall required a mental readjustment, and for some people this meant distancing themselves from what they had declared a short time previously. Ten days after the informal Council at the Elysée, on 28 November 1989, Kohl presented his famous ten-point plan for German reunification. This encouraged everyone at home and abroad to get a move on. Crucially, Kohl wanted to embed German reunification within a united Europe. This concurred with his deepest convictions. He wanted to strengthen and broaden the Community with all that this implied at the time: the EMU and the Social Charter and, further in the future, the European Political Union (EPU).

German Reunification

Events in Central and Eastern Europe and in the Soviet Union followed one another in quick succession. Because the situation was changing from day to day and because there was so much at stake – both in the context and the substance of European unification – bilateral contacts took place continuously, principally by telephone. The run-up to the European Council meeting of Strasbourg was characterised by intense diplomatic exchanges.

The summit on 8–9 December 1989 began with an account by Delors of the implementation of the Single European Act and its influence on economic growth. During lunch the evolution in Central and Eastern Europe was discussed. The afternoon session was dominated by the EMU dossier. The debate on the Social Charter was a lot more difficult: it was abundantly clear that Margaret Thatcher would never agree.

The most important and most delicate passage in the conclusions had to do with German reunification: "Our aim is to perpetuate the state of peace in Europe, in connection with which the German nation will regain its unity by free self-determination. This process should be carried out in a peaceful and democratic way, taking into consideration agreements and treaties and all the principles laid down in the Helsinki Final Act, in a context of dialogue and cooperation between East and West. At the same time, we must not lose sight of European integration". In other words: the principle of "not a German Europe, but a European Germany" had been attained. But we certainly had our work cut out for us.

In the dinner that preceded the final outcome, Mitterrand did not take the same line as Kohl. He was very cautious and he still did not formally declare himself in favour of reunification. He would not do so until much later. After Strasbourg, to the great displeasure of Kohl, Mitterrand was to make one more official state visit to the GDR on 20–22 December 1989. Later Mitterrand attempted to defend himself against this criticism by referring in a book to my official visit to Poland in April 1991. Although I had made this visit to Poland before the communist regime had finally disappeared, it was after a political "round table" had been held with the decisive participation of Lech Wałęsa and his Solidarity movement.

During the dinner in Strasbourg, German reunification was already a *fait accompli* in Kohl's mind. He was anxious to convince the opponents and doubters among his fellow government leaders. Thatcher declared herself against the reunification. Mitterrand hesitated, so did Andreotti. Gonzalez, Santer and I were strongly in favour. Lubbers intervened in the form of a question: "On the basis of the past, is it opportune for Germany to become united again?"

This *tour de la table* left deep scars. Kohl wanted to force a breakthrough and not everyone appreciated it. He was furious with Lubbers' intervention. As he left the dinner Kohl snarled at Lubbers: "I will teach you something about German history!" For Kohl, the history of Germany is part of his very being. His elder brother, Walter, died in 1944. His memories of the Second World War – he was too young to have taken part, too old not to know about it – are at the very core of his personality and his political commitment. Hence, he felt that he did not need any lessons in German history from the Dutch Prime Minister. Lubbers came to regret this incident.

Although the two of them still got along well together thereafter, a nerve had been touched. On the other hand, strong Spanish support for unification was based on the personal friendship between Kohl and Gonzalez. This would do Gonzalez no harm. It plays a substantial role in Kohl's memory.

German reunification, a social Europe – which was treated harshly by the Single European Act and, because of the abiding opposition of the British, never got off the ground – and the necessity for accelerated integration continued to dominate debate, even after the Strasbourg Summit, but contradictory scenarios circulated about the why, how and when. The geopolitical situation did nothing to contribute to stability. It was the period for many of the speculations on the future organisation of "the Common European House", as Mikhail Gorbachov had called it six months earlier.

Towards a Political Union

At the end of 1989 in Strasbourg there was still a great deal of uncertainty, but during the Irish Presidency in the first half of 1990 the German question was given the green light. The realisation that in the light of the impending reunification of Germany new steps also needed to be taken towards political unity had made itself so keenly felt that the ministers of Foreign Affairs were given the task of preparing an amendment to the treaty.

The Italian Presidency was to point the way. At the European Council of Rome in December 1990, a decision would be made to hold two Intergovernmental Conferences, one on the EMU and one on political union. Delors was anxious not to overdo things, because he feared that then nothing would ultimately come of the single currency. There was a very real danger that the momentum would be lost. The French wanted to speed up the implementation of the single currency, therefore, because Paris feared that the Germans might change their minds. But, at the same time, the conviction had gained ground that only a strengthened Community could play a role in the new Europe – hence, the necessity for a second Intergovernmental Conference on Political Union.

Meanwhile, France also brought up the question of the site of Parliament again. The reunification of Germany had made Strasbourg more than ever the symbol of Franco-German reconciliation and of the Paris–Bonn axis in Europe. But Brussels also gained importance from the strengthening of the Community. My position was that the Member States had also to take into account the wishes of Parliament. We were nowhere near a solution, and the Italian Presidency was given the task of working out a definitive ruling by the following European Council.

Twice in Rome

The fear of losing momentum proved to be real. On 2 August 1990, Iraqi troops entered Kuwait. The worldwide boycott of Saddam Hussein's regime and the war in the Gulf region shattered optimism. The economic recession also affected Western Europe, and Germany had to dig deep into its own pockets to pay for reunification. The divided and powerless response of the EU during the war in Yugoslavia did nothing for its reputation. This gradually affected public support for the European project and two years later this would lead to a crisis over the ratification of the Maastricht Treaty.

By the end of 1990, it was the Italians' turn. Just as in the preparation for the Single European Act, Council President Andreotti played a star role. During a meeting on 16 September 1990 in Cagliari, he swore to me that the EMU would now definitely be pushed through with German support. If not, the EMU would most likely have become improbable a year later. For Andreotti the complete transfer of Parliament to Brussels was impossible, though he did consider a status quo or a temporary ruling feasible. I should not have expected too much help from the Italians, as later became apparent. But Andreotti's handling of the EMU and the European Political Union (EPU) was faultless. As EPP President, I fully supported him. The fact that an Italian Christian Democrat was leading the Community increased the chances that our ideas would actually become reality.

The first European Council after German reunification took place in Rome on 27-28 October 1990. I remember the heated discussions between Thatcher and Andreotti over the single currency. Andreotti succeeded in completely isolating Thatcher and making her swallow the EMU despite her loud protests. She was given just one paragraph in which she explained the dissenting British position. It immediately became the swansong of the Iron Lady. The longer she held on to her Eurosceptic attitude, the more counterproductive it became, and she started to feel the pinch even on the home front at that point. On 27 November her own party forced her to resign and she was succeeded by John Major.

The second European Council in Rome, on 14-15 December, set in motion a treaty which would later take the name Maastricht. Enlargement was pushed ahead, but first the two IGCs had to reach conclusions, by the end of 1991 at the latest. Ratification had to be complete by the end of 1992 so that the new treaty could come into force on 1 January 1993. After the summit had ended the two IGCs were formally opened. At this European Council, the first since 1979 without Mrs Thatcher, I was now the government leader with the greatest longevity. There were great hopes that the attitude of the United Kingdom would now be reversed. A new period was heralded without negative, sceptical viewpoints. It was even predicted that Major would accept the euro and the Social Charter of the Treaty. His first interventions did indeed feed that hope. But only too soon Major would fall back on some of Thatcher's viewpoints. That was not yet clear in Rome, but would be apparent not long afterwards.

A Role for the EPP

The Luxembourg Presidency in the first half of 1991 introduced the double IGC as a permanent feature. It was too late for simply spouting ideas; a new treaty now needed to be written. The EPP also provided the impetus for that preparation. Our contribution, as well as that of the team of Mitterrand, Kohl and Delors, was crucial and decisive for what later became the Maastricht Treaty. The European Council consisted of twelve members at that time. There were two Socialists, François Mitterrand and Felipe Gonzalez; six members of the EPP, Helmut Kohl, Ruud Lubbers, Jacques Santer, Constantine Mitsotakis, Giulio Andreotti and I; the British Conservative John Major, the Danish Conservative Poul Schlüter and the Portuguese Anibal Cavaco Silva (at that time a member of the Liberals who would later join the EPP); as well as the Irish Fianna Fàil leader Charles J. Haughey.

We were aware of the immense challenge and also the great responsibility involved. It was now or never. The EPP Summit met twice under my presidency: on 13 April in Brussels and on 21 June in Luxembourg. The first meeting was still dominated by the question of the membership of the British Conservatives in the EPP. The second, at Senningen Castle, prepared for the European Council of 28-29 June, which brought Luxembourg Presidency to an end.

Positions were adopted, alliances forged, negotiating margins explored. Kohl, for instance, floated the idea of a congress of European and national members of parliament, because this new institution would provide an excuse for not having to extend the powers of the European Parliament. Even the timetable was tinkered with. The original plan to come up with a new treaty during the Luxembourg Presidency was abandoned, since it was not feasible. We gave ourselves six months' extra time to get everyone into line, because ideas about political union still appeared to be insufficiently developed. As far as Kohl was concerned, there could be no question of an EMU unless the results of the EPU were at the same "level". Even though postponement always carries a risk, Kohl pointed out that the Dutch Presidency would be led by Christian Democrats, specifically Lubbers and his Minister of Foreign Affairs, Hans van den Broek. That gave him hope of a good outcome, for Germany, for Europe and for the EPP. The fear arose that the new treaty would not bring what many hoped for. Two years earlier, in Strasbourg, the bar had been set very high, but on the course to Maastricht there was the temptation to lower ambitions.

The EPP once again drove the outcome of the IGCs during the Dutch Presidency. I was in charge of a working group of staff of the EPP government leaders. In addition, the EPP Summit met three times: on the fringes of the NATO summit in Rome on 8 November 1991; at Stuyvenberg Castle in Brussels on 26 November and in The Hague on 6 December. There I was also designated as EPP spokesperson to introduce the treaty article on the recognition of European political parties during the European Council.

I will use an example to illustrate the fact that the EPP summits really did make a difference. Lubbers was not greatly in favour of augmenting the powers of the European Parliament and he felt the same way about the *avis conforme*. At the EPP Summit of 26 November, he had to backtrack as Council President if he wanted to retain the support of the other EPP prime ministers. However, the *avis conforme* did not survive during the final negotiations.

At the Maastricht Summit I was pleased about the EMU but not about the EPU, even though significant successes were achieved. Once again the British had put a spike in the wheel, and at a crucial moment too. Maastricht was a political agreement and a concession. The combined goal of a European political and monetary union, which I and the other EPP government leaders dreamed of, was for some leaders too much to swallow – what a pity.

Among Christian Democrats

In spite of its shortcomings, Maastricht was Lubbers' moment of glory. It was partly due to him that the treaty was finally concluded. I had known Lubbers for a whole decade by then. Our first meeting dates from the period when he was CDA party leader in the Dutch Parliament. Dries van Agt was still Prime Minister at the time; I was already Prime Minister. We saw each other frequently at meetings of the EPP as well as at European Council meetings. We were also in regular contact between meetings. From the start something had clicked between us, particularly on a personal level, even though there were obvious differences. I still regard him as a close friend. In terms of policy, we had some differences in emphasis. He was less Benelux-oriented. Like so many Dutch politicians, he was closely attuned to London. "That is my role", he told me. "That is what I must do". He frequently put himself in the line of fire and was probably the only European leader who maintained good contacts with Margaret Thatcher – he was often given the task of persuading her.

His relationship with Kohl was very much an ambivalent one. Kohl relied on Lubbers to a great extent as a Christian Democrat who was positively disposed towards Europe. This was repeatedly apparent, including at Maastricht. But during the discussions about German reunification they clashed head-on. Kohl never forgot that. This was apparent when Lubbers wanted to become Commission President in 1994, and also when he stood as a candidate for Secretary-General of NATO. Each time he had the support of the British, but he never made it. In 1998, Lubbers successfully became High Commissioner for Refugees at the United Nations in Geneva. He had finally got an international position, but now his extrovert character proved fatal. He embraces everyone. He has always been like this as far as I can remember. In Geneva he was made to pay for this but I, nevertheless, remain firmly convinced of his integrity.

The same was true of Andreotti, whose reputation was squandered for other reasons. He was pursued by prosecutors for many years, but he was always acquitted. His contribution to the success of the IGCs on the internal market and the euro is indisputable. He knew what he was talking about, because he had been involved in Italian politics since the 1970s, as minister and Prime Minister. Andreotti is extremely intelligent, a convinced European, inventive, mysterious and intriguing. It is rumoured that he gets up at four O'clock every morning, goes to Mass and never stops writing. I can confirm the latter. He used to write continuously during the meetings as a member of our parliamentary group in Strasbourg. In spite of his advanced age, he is still active in Italian politics and is often quoted in the media. He is a true survivor.

My Battle for Brussels

It was only over the site of the European Parliament that Andreotti did not fully support me, but then, neither did Lubbers. As I have already mentioned, I opposed its definitive location in Strasbourg for the first time in 1981. I have always defended the joint functioning of the Commission, the Parliament and the Council in Brussels as an absolute necessity for achieving proper synergy and functionality in the EEC/EU, not to mention the ongoing cost of relocating to Strasbourg one week every month. I continued to plead for a different but significant role for Strasbourg. But that was not what the French wanted to hear.

In Dublin, it had been decided that the Italian Presidency must work out a definitive proposal by October 1990. On 17 September, I met a delegation of Belgian MEPs who informed me of a fierce anti-Belgian attitude in Strasbourg. In response I let it be known that I would use my right of veto if the status of Brussels came under threat. But the Italians did little or nothing that was likely to contribute to a solution. The fact that Andreotti and his Minister of Foreign Affairs, Gianni de Michelis, had openly expressed their preference for Strasbourg did not exactly help matters to progress. The passage on the site of Parliament was deleted from the resolutions and the matter did not even appear on the agenda.

At the second European Council in Rome in December, a first attempt was made to reach broader agreement on the location. It was proposed to establish the Commission and the Council in Brussels, with the exception of the Council sessions in April, June and October in Luxembourg, and the Parliament would meet in Strasbourg for all the plenary sessions. The committees, the working groups and the political groups would meet in Brussels.

I refused and no agreement materialised. Just before the Rome Summit, I was supported by Parliament, which pleaded in a resolution for a single site for Parliament, Council and Commission. Mitterrand was not exactly overjoyed by my stubborn attitude. After the summit, he remarked during his press conference that if Strasbourg was challenged as the location of the Parliament, then why not also challenge Brussels as the location of the Commission? Belligerently, he added that he was in favour of grouping the institutions together, but if this was not to be in Strasbourg, then it should not in Brussels either. He announced that he would continue to defend Strasbourg, yet without maintaining that it was the capital of Europe.

The French campaign certainly persisted. I therefore once more received Elisabeth Guigou, the French Minister of European Affairs, at Stuyvenberg castle on 30 May 1991. There was also a rumour that I was prepared to waive my opposition in exchange for a high European position, particularly the presidency of the Commission. The rumour that I would buy my way into the presidency by abandoning Brussels as the site of the Parliament was absolutely scandalous. I never considered such a thing. If I wanted to fight for Brussels as the capital of Europe I had to be a free man and could not foster any personal ambitions.

The question of location remained unresolved and was passed on to the Dutch Presidency. Lubbers had worked out a system whereby the location would be definitively established in Strasbourg as long as a few concessions were made to Brussels. Lubbers had suggested this to Mitterrand in a letter. He had some kind of horse-trading in mind, whereby he angled for French support for the candidacy of Amsterdam as the location for the proposal European Central Bank. Whether Mitterrand would have supported Amsterdam against Frankfurt is very doubtful. When the letter became public, the compromise was dead in the water. Anyway, it was unacceptable to Belgium for the President of the Council to work out an agreement without consulting Brussels. I stuck to the status quo of Maastricht, insisting that the transitional measure of 1981 would last and that the wishes of Parliament be respected.

At the Maastricht Summit, I refused to discuss Lubbers' compromise solution, which had already met with Mitterrand's approval. Kohl followed the French. Other government leaders were in favour of Brussels; some stuck to the status quo or did not take a view. In Lubbers' proposal, the Parliament would continue to hold its normal monthly sessions in Strasbourg. Other plenary sessions would be held in Brussels, as would the committees and the political group meetings – this was regarded as a *fait accompli*. I was playing for high stakes by demanding recognition of Brussels as the definitive home of the European Parliament. This put a roadblock in the way of the whole location debate. Diplomats were surprised at the obstinacy with which I stuck to my view and the efficiency of my strategy. By playing for such high stakes, Brussels might well lose everything.

Lubbers' proposal was ultimately upheld in the solution that was adopted at the Edinburgh Summit on 11–12 December 1992. Since then, twelve plenary sessions a year have taken place in Strasbourg, including the budget session. Extra plenary sessions, as well as the meetings of the parliamentary committees and the political groups, are held in Brussels. Most of the Secretariat is still located in Luxembourg. In principle the French got their way, but in practice everything happens in Brussels. The fact that my successor as Prime Minister of Belgium, Jean-Luc Dehaene, took an active part in fashioning a broad agreement, in which Brussels gave up its ambition to become the definitive home of Parliament, certainly facilitated his candidacy for the presidency of the Commission. I am convinced that I sowed what Dehaene was then able to reap. Without my rigid attitude we would definitely have been trampled underfoot.

Mid-Term Review

In spite of the rocky road to Maastricht, the period 1989-1991 can be regarded as extremely successful for the EPP. The Maastricht Treaty owes a great deal to the suggestions and ideas of the EPP, resulting as it did from the indefatigable and unanimous efforts of heads of government affiliated with the EPP to create a federal Europe. Most of the objectives we had laid down during our EPP Summit, in The Hague on 6 December 1991, have been met: the simultaneous terms for Parliament and Commission; the vote of confidence in the Commission by Parliament; the expansion of Community powers; the joint decision procedure; the EMU and an independent European Central Bank; an advisory Committee of the Regions; a massive extension of majority voting; recognition of European political parties and so on.

Some objectives have not been met, such as the formation of a single community, the *avis conforme* and the phasing out of unanimity on foreign policy. The new European Union still does not have a federal structure, in spite of the fact that the principle of subsidiarity was acknowledged.

The Maastricht Summit brought an end to what was for me a long series of European Council meetings. The ceremony on 7 February 1992, at which the ministers of Foreign Affairs and Finance signed the new treaty, was my last appearance abroad as Prime Minister of Belgium. In spite of the new treaty, the problems of Europe were far from solved. Ratification was not entirely smooth sailing, and the ratification of the Maastricht Treaty itself became a problem. On top of that, since Maastricht, the foundations of European integration have been challenged. In particular the negative campaigns during the referenda have brought to the forefront the discussions about the democratic deficit in the Union. Even though I was no longer a member of the European Council, I still followed the European dossiers as EPP President and could not get them out of my head. So I maintained my commitment to Europe and I remained closely involved with the integration of a united Europe.

Chapter VI

"O Hellenic Shore Where Our Fathers Dwell …"

I have never lost sight of the European dimension in my political activities. In fact, my commitment within the EPP gathered increasing momentum in the early 1990s. Only a few weeks before I was to meet my fellow heads of government at the EPP Summit in Pisa on 17 February 1990 – during which Helmut Kohl succeeded in convincing Giulio Andreotti to support German reunification – Thomas Jansen inquired discreetly as to whether I might be interested in becoming President of the EPP.

Jansen, a German, had been Secretary-General since 1983. His question stemmed from the good relations and cooperation we had built up over the years. A conflict between Jansen and Kohl, which had been dragging on for some time, also played a role. Helmut Kohl had promoted Jansen to Secretary-General but during the late 1980s had become increasingly dissatisfied with the state of affairs within the EPP. Kohl had decided, therefore, that the CDU would no longer pay its membership dues. This would have been fatal for the operation and credibility of the EPP. Jansen hoped that if I were President, given the fact that I was on such friendly terms with Bonn, relations between the CDU and EPP would become normal again, and that the practical problems regarding the cost of membership would be solved.

I agreed, because I believed that Jansen had asked me on behalf of the CDU. However, during the Pisa Summit I was approached by Kohl, who said, "I hear you have agreed to stand for President. What is going on? How come I was not informed? Nobody told me". It was not unreasonable to assume that Kohl had already been informed of the procedures that would lead to the election of a new President. My presidency seemed to be getting off to a bad start. But the misunderstanding was soon cleared up. Fortunately, we had known each other for years and our mutual understanding was not at all damaged by the incident. On the contrary, Kohl was very much in favour of my candidacy and has supported me as President to this very day.

Rekindling an Old Flame

In 1990 the EPP President was still elected by the EPP Political Bureau. This comprised a quite limited number of representatives of member parties and members of our group in Parliament. The EPP had only ten member parties at the time: the original nine plus Nea Demokratia, which became a full member following Greece's entry into the Community in 1981. I was elected President on 10 May by a vote of sixty-five to two. There were no abstentions.

I remember very well having lunch with Leo Tindemans and Jacques Santer after the election. Tindemans was an MEP at the time and he was not all too happy with my election because he had been asked to resign as President three years earlier following some slight pressure from Kohl. I was taking over from Jacques Santer who had presided over the EPP since 1987. As he was Prime Minister and lived in Luxembourg he was unable to dedicate himself to the position. This caused considerable dissatisfaction as a result. He was accused of lacking initiative, for example. During the EPP Congress in Luxembourg in 1988, Kohl made some extremely critical remarks to the heads of government present about how badly the congress was run. But the dissatisfaction was even more deeply rooted. There were problems in relations between the Party and our group in the European Parliament; cooperation between the individual parties was poor and, as a result, the Group suffered heavy defeat in the 1989 European elections.

In fact, since the foundation of the EPP in 1976, the role of President had been merely ceremonial. It could be said that the presidency was held by someone who had been "put out to pasture", so to speak. The statutes limited the President's duties to chairing congresses, meetings of the Political Bureau and EPP summits. In fact, the EPP was run by the Secretary-General because he or she was available on a permanent basis and lived in Brussels.

When I became EPP President in 1990, it was far from obvious that I could be actively involved, as I was still Prime Minister of Belgium at the time. But because I lived in Brussels on a permanent basis, it was easier for me to take up both positions. I changed the calendar of events in the government from time to time in order to be able to conduct EPP business, attend EPP Congresses or go on official visits abroad. Jean-Luc Dehaene, my Vice-Premier at the time, once said: "The Prime Minister is always busy with Europe!" Only after my resignation as Prime Minister in March 1992 did a definitive change take place. The EPP presidency became my full-time job and I transformed it into a truly political position.

My First Congress

The EPP Congress in Dublin on 15–16 November 1990 was my baptism by fire. My host was Alan Dukes, leader of Fine Gael. This was the first congress after the fall of the Berlin Wall, the revolutions in Central Europe and Russia and the reunification of Germany. The congress was held in between the two European Councils of Rome and just before the start of the double Intergovernmental Conference on the EMU and the EPU. This was hardly unimportant, as the EPP planned to exert considerable pressure regarding the content of the negotiations.

The programme and new statutes were drawn up in the run-up to the congress. I was only indirectly involved, as I had assumed the position of President only in May and proceedings had already been initiated by Secretary-General Thomas Jansen. The main point for me was the congress itself. It formed a tremendous challenge, which, in retrospect, I had nonetheless underestimated. Similar to the process at national congresses, discussions were first held in working groups. Amendments were tabled and then the debate was re-held during plenary sessions, where a vote was cast on the document, "A Federal Constitution for Europe". Both the title and the text were visionary and ambitious, and it is very much a point of discussion whether the project is still at all possible today.

The purpose of the new statutes was to make the EPP function more efficiently. The initial subtitle of our name, Federation of Christian Democratic Parties of the European Community, was simplified to Christian Democrats. Parties from candidate Member States could become associate members while awaiting full membership. Parties whose political leanings were close to those of the EPP but whose countries were yet not candidate Member States were to become observer members. Individual membership was also established. The Conference of Government and Party Leaders, later the EPP Summit, became an official Party body. I was able to make a fresh start with these newly adopted EPP structures.

Celtic Warriors

I got to know the Republic of Ireland through my colleagues as heads of government, Charles Haughey and Garret Fitzgerald. I had considerable contact with Fitzgerald and his successor John Bruton, since Fine Gael was a member party of the EPP. They were there at the very beginning when the EPP was founded. As a centre party with a strong focus on European integration, Fine Gael fit wonderfully well into our political family. Compared to other parties, Fine Gael had undergone a considerable number of political changes in its history. It has a strongly democratic culture in which important subjects are discussed in great depth. It was impossible for the other large Irish party, Fianna Fàil, to join the EPP. Historically, however, Fine Gael was their main opponent. Nearly all of the parties with which Fianna Fàil had formed a group in Irish Parliament have become members of the EPP.

The Ireland I got to know in the 1980s was not unlike the Flanders I had grown up in: poor, rural and Catholic. Since then the country has undergone a major revolution, becoming one of the most prosperous regions in Europe, thanks to joining the EEC in 1973. This was not always appreciated by the population, as became clear in 2001 and in 2008 when they rejected the Treaty of Nice and the Treaty of Lisbon in their respective referenda. Be that as it may, Ireland still forms convincing proof that the EU can be decisive in transforming a country.

The recent success in Ireland has strengthened its orientation towards the continent rather than towards Britain. There is nothing strange in this, given its troubled history with Eurosceptic Britain and its familiarity with Catholic culture on the European continent. At least that is the view of John Bruton, who is now EU ambassador to the US. As a "European", he feels more at home on the continent than in a predominantly Protestant environment, he once said to me. This is hardly surprising, as I know John to be one of the most European-minded Prime Ministers ever and also a very active and dynamic Vice-President of the EPP. What he fought for as chair of the Athens Group – the preservation of the Christian Democratic roots of an ever-broadening EPP – is in complete agreement with my deepest convictions.

Deepening and Widening

The Dublin Congress provided the central ideas for deepening the Party, but the real breakthrough would occur only later. In my opening speech, I pointed out that "If the Union is to form the cornerstone of a future greater Europe, this cannot be achieved by looking inwards or by allowing itself to become watered down in some vast construct before achieving its goal. Only through the strengthening of the links between our twelve countries will it become possible to expand and include all other democratic countries that are willing to join the European Union". The further development of a European party was already central to the discussion. I therefore closed the congress with the following words: "The 'new world' that is emerging can only be really new if it inspired by 'new thinking'. It is the determinative, irreplaceable role of this political party to develop and defend a model of society and to convince people of its worth".

The expansion of the Party was not at the top of the agenda in Dublin, but we did get a taste of what was to come. A delegation from Central and Eastern Europe attended the EPP Congress for the first time. The congress was addressed by one of their representatives, Alojz Peterle, leader of the Christian Democrats and Prime Minister of Slovenia. This was quite striking as Slovenia had achieved independence only on 25 June 1991.

The British Conservative Party also sent a delegation, led by Sir Christopher Prout, chair of the European Democratic Group in the Parliament. Prout was on good terms with our EPP Group and also with its chair, Egon Klepsch. Given the prominence of the Group at the congress, the British press had turned out in force.

In addition to this, there was a power struggle going on at the same time involving Margaret Thatcher.

There was a delegation present from the Spanish Partido Popular (PP), including the then unknown José María Aznar. His attempts to approach the EPP were hindered at the time by Christian Democrats from the various regions of Spain. Out of protest, the President of the Basque Partido Nacionalista Vasco (PNV), Xavier Arzalluz, addressed the congress in German. The purpose was to deliberately provoke Aznar. I reacted immediately to this barefaced attack and succeeded in getting Aznar back to the table. I felt intuitively that Aznar should be supported, despite opposition from Spanish regional Christian Democrats. For this reason alone I felt it important to welcome Aznar and to give him his deserved place at the congress. My experience with Spanish people had taught me that such matters were highly sensitive. Had I not given the appropriate respect to the leader, our contacts with the PP would have ended in disaster.

Opting for Aznar

The accession of Spain and Portugal to the European Community in 1986 had weakened the EPP from an electoral point of view, because these countries did not have strong Christian Democratic parties. The Christian Democratic party Centro Democrático Social (CDS) had been falling behind in Portugal, whilst in Spain there was only one small national Christian Democratic party: the Partido Demócrata Popular (PDP). This party was more or less the successor of the pre-war Spanish Partido Social Popular, which had been set up in 1922 through the Sécretariat International des partis démocratiques d'inspiration chrétienne (SIPDIC) and had maintained contacts with the other Christian Democratic parties in Europe.

Roberto Papini has described this in detail in his work *The Christian Democrat International* (Papini, 1977), as well as in his book about Luigi Sturzo and the international organisations of people's parties between the two world wars. This was translated by Alain de Brouwer and published under the title *Le courage de la démocratie – Sturzo et l'Internationale populaire entre les deux guerres. Matériaux pour une histoire* (Desclée de Brouwer, 2003).

Don Luigi Sturzo was a priest who founded the Partito Popolare Italiano (PPI) in 1919, the first people's party in Italy. The party was banned by Mussolini in 1926. Its leaders and activists went into hiding or into exile. Those who had gone into hiding joined the resistance during the Second World War and, led by Alcide de Gasperi, they re-formed the party after liberation in 1944, which they then called the Democrazia Cristiana.

Don Luigi Sturzo escaped to London. He inspired and organised international cooperation between democratic parties based on Christian ethics. In fact, during his exile between the two world wars he went on to breathe life into European Christian Democracy and to become one of its founding fathers.

The Christian Democrats in the various regions of Spain – the predecessors of the PNV in the Basque Country and the Unió Democràtica de Catalunya (UDC) in Catalonia – had worked together with Don Luigi Sturzo and his international secretariat until the outbreak of the Spanish Civil War put an end to their cooperation. Only after the collapse of Franco's regime and the reinstatement of democracy did these parties re-emerge. They had considerable support in the regions but were not very prominent at the national level.

The EPP had tried for a long time to set up a national Christian Democratic party in Spain, at first immediately after the fall of Franco and afterwards when Spain joined the European Community. In the mid-1970s I had already met Christian Democrats such as Ruiz Giménez, Javier Rupérez and José María Gil-Robles senior, the father of José María Gil-Robles, who later became the President of the European Parliament. Because their parties were members of the EUCD, they were not unknown to us. The new party promoted by the EPP, which would bring together national and regional Christian Democratic parties, already had a name: Unión de Centro Democrático. But it never succeeded in coming together. The various small Christian Democratic parties continued to exist alongside each other. For this reason, the EPP did not put all its eggs in one basket, as it were. We had already established contacts with other parties. I even remember negotiations with Jordi Pujol, the Catalan nationalist, and with Manuel Fraga Iribarne, leader of the Alianza Popular, who had been a minister under Franco.

Fraga's party was much stronger nationally than were the Christian Democrats. The Alianza Popular sought to become a member of the EPP Group in the European Parliament because it no longer wished to form a group with the British Conservatives. A crucial factor in their application for membership in the EPP was the formation of a new political entity, the Partido Popular, at the beginning of 1989 as the result of a merger between the conservative Alianza Popular, the Christian Democratic PDP (which had re-named itself Democracia Cristiana in 1988) and a smaller Liberal Party. A breakthrough between the Partido Popular and the EPP Group was reached in the spring of 1989. Essentially for this purpose, and also for subsequent membership in our group, we developed a sophisticated formula: MEPs who were elected under a Christian Democrat flag could join the EPP Group. Because PDP Christian Democrats were candidates on Partido Popular's electoral list, the entire list could become members of the EPP Group. Indeed, this happened after the 1989 European elections.

In April 1990, Aznar succeeded Fraga as President of the Partido Popular. This change in leadership formed a "point of no return" in renewing centre-right politics in Spain. Aznar got in touch with the EPP immediately following his election. His goal was to remove himself from national and international isolation as quickly as possible by setting up closer ties with the EPP. As EPP President,

I met him on 16 June 1990 in Brussels in the presence of Javier Rupérez, the last chair of the PDP. Marcelino Oreja, the liaison between the Alianza Popular and the Spanish Christian Democrats, was not present at the meeting, something which he deeply regretted. Oreja was the "John the Baptist" of the Partido Popular.

In light of what was to come, this meeting proved to be extremely important. "Your ambition should be to become Prime Minister of Spain", I told Aznar at the time. "You have to be able to provide an alternative to the left and we are going to work together on it". Aznar came under severe attack within his own party for approaching the EPP, not the least from former Alianza Popular conservatives. At the same time, there was strong opposition from within the EPP as well.

When it was proposed during the meeting of the Political Bureau on 4 October 1990 to give the Partido Popular observer status, not only the regional Christian Democrats from the Basque Country and from Catalonia voted against it, but also the Dutch CDA and the Italian Democrazia Cristiana. Aznar was mainly accused of not being a "true" Christian Democrat, and the same critique was aimed at the Partido Popular. This was true to some extent but the real problem lay elsewhere. The EPP asked itself the fundamental question: Did the Partido Popular really belong among its ranks? Though they had different origins and traditions, the EPP and the Partido Popular were both people's parties in name and in purpose. More than that, they were natural allies. Admission would become a fact as soon as Aznar embraced our values and principles. And this he did, which was far from obvious, given the prior history of the Partido Popular. His declaration during the EPP Congress in Dublin that he would uphold and promote Christian Democratic values in Spain caused uproar in the Spanish press. Aznar too had taken considerable risks.

Because relations between the Partido Popular and the EPP were excellent, and because Aznar wished to root his party firmly and permanently in the EPP, he expressed the wish to become a full member of the Party. This occurred during the opening of the EPP Group workshops in Santiago de Compostela. His request for membership had taken place in agreement with me – I thought that the time was ripe to do so. The EPP Political Bureau, which was held on 18 October 1991, passed the motion unanimously, with the exception of the Basque PNV, who voted against. Both the representatives of the PNV and its chair, Arzalluz, continued to oppose the Partido Popular within the EPP. This led in 1999 to an unavoidable split.

No Reservations

The Partido Popular's membership in the EPP proved a success as well as a fine example of a win–win situation. The EPP was now strongly represented in Spain and the voice of the Partido Popular could also be heard at the highest levels in Europe. Moreover, Aznar continued to plead for the broadening of the EPP, by supporting the inclusion of Silvio Berlusconi's Forza Italia, for example. Moreover, he had the great intelligence to steer his party towards the centre of the political landscape in Spain. He also rejuvenated his party considerably and brought in more female representatives. He had already achieved considerable success in the 1994 European elections, almost doubling the number of Partido Popular seats. In 1996 he went on to win the national elections and obtained an absolute majority in 2000.

I have always got on well with José María Aznar on a personal level. Ever since our first encounter, we have shared an increasing degree of mutual trust. Our conversations have always been open and honest, despite our personal differences. Aznar is radical in his thinking and seldom wavers from his own point of view, even in its details, whereas I am more inclined to search for a consensus. There is seldom room for compromise, as far as he is concerned. In other words, it is quite difficult to get him to move on a point. That can only be achieved with considerable patience and the power of persuasion, which we were to discover on a number of occasions. His stance has much to do with the presidential management style within the Partido Popular. Loyalty to the leader is extremely strong within the party, and cooperation is next to impossible if this is not taken into account. If I had not done so, our good relations and high levels of understanding would also have proved impossible.

Personally, I have never hesitated to support him politically, in contrast to other heads of government within the EPP. It often happened that others failed to show their true colours. I continued to pursue this line as President of the EPP and attended practically all of the Partido Popular's election campaigns during their period of growth, invariably delivering my speeches in Spanish. José María was elected Vice-President of the EPP in the spring of 1993. Afterwards, I held a well-attended press conference during which I introduced him to Europe for the first time, as it were. We held our congress in Madrid on 5 November 1995. In my opening speech I made my intentions quite clear from the outset: "We have come to Madrid to show our support for the Spanish People's Party and its President, José María Aznar, and we are confident of his forthcoming victory in the elections .... José María Aznar, we stand shoulder to shoulder with you and your supporters in the Spanish People's Party".

The congress was marred by a serious incident. PNV chair Arzalluz declared, much to everyone's dismay – and not in German this time – that he had no intention of supporting Aznar as Prime Minister. In an impromptu declaration, Arzalluz dictated to us what we all should do. I said that it was the duty of all Christian Democrats who were members of the EPP to stand behind Aznar after the elections in his attempt to form a majority. My intervention was broadcast repeatedly on Spanish television. People even recognised me in the Madrid metro. Perhaps that was my most efficient form of support for José María Aznar.

Stalwarts

Little by little the Partido Popular became one of the stalwarts of the EPP. It was the most important party next to the German CDU/CSU. The Partido Popular filled the vacuum left by the implosion of the Italian Christian Democrats. In early 1999, when the position of EPP Secretary-General became vacant, at a time when the CDU had their hands more than full with their financing scandal, the Partido Popular insisted that the position be given to them. They had a right to it, given the size and importance of their party. This is just one of the many forms of balance that we in the EPP, and I as its President, have continually to keep in mind.

And so Alejandro Agag Longo, Aznar's personal secretary, became Secretary-General of the EPP in 1999. His election to the post would have proved impossible were it not for the good relations between Aznar and myself. Agag also strengthened our relations at the same time. He had already played an important role in preparing and organising the 1995 EPP Congress in Madrid. It was also thanks to him that the congress was a success for the Partido Popular. Afterwards, I regretted being in favour of his election as MEP. In a way, I was alone and was forced to be active full time within the EPP. It would have been more normal for the President to be an MEP and the Secretary-General to take care of the practical running of the EPP. Agag's departure from politics in 2002, much against Aznar's wishes, nonetheless came as a surprise. It was a loss for the EPP and for the Partido Popular. He could still have played an important role in national or in European politics. Agag was succeeded by Antonio López-Istúriz, who was a personal secretary to Aznar and remains Secretary-General to this day.

My open support for Aznar to become Prime Minister in the early 1990s was not without results. The then Spanish Prime Minister had already confided in me *in tempore non suspecto*: "I have one great problem in Spain. We have no alternative". Of course, the arrival of Aznar changed all that, and my relationship with Gonzalez as well. I could no longer rely on his support when my name was put forward to be President of the European Commission. He once said that he would have had supported me to the very end, until the day I spoke in favour of Aznar as Prime Minister in Spain. At no time have Gonzales and I ever been enemies, however. Good relations were soon restored between us.

Like Ireland, Spain is a clear example of success as far as European integration is concerned. Next to its economic growth, its definitive embracing of democracy is also very clearly visible. The fact that political parties play an essential role in the democratic process has been enshrined in their constitution. Though its culture is Mediterranean, there is an enormous difference between it and Italy, for example. Political parties are well structured and organised to an important extent around the personality of the party leader, who automatically becomes Prime Minister if the party wins the elections. This "presidential" system is founded on enormous internal cohesion. Division within a party is abhorred. Despite the strong sense of nationalism in Spain, it is one of the most pro-European Member States in the Union. This became clear during the referendum, when Spaniards voted in favour of the European Constitutional Treaty. It is also clear from the stance of many of its politicians, like the late Loyola de Palacio, for example. Like Aznar, her political roots lay in the Alianza Popular. But her background and ideas did not push her in the direction of Euroscepticism. The same is true for the new leader of the Partido Popular, Mariano Rajoy, who has continued to carry forward José María's political heritage in difficult times and with success. I had very much hoped that Loyola would play a crucial role in the EPP, following her vice-presidency of the European Commission. She would have been a brilliant President for the EPP. I was deeply saddened by her untimely death and, also sadly, along with her my dream for the EPP passed away as well.

Back to the Barricades

When Jean-Luc Dehaene succeeded me as Prime Minister in March 1992, I moved from 16, rue de la Loi to the headquarters of the EPP in 16, rue de la Victoire. I could take refuge there as President of the EPP: a small office, three metres by four. My arrival brought about a complete change. We held press conferences involving twenty to thirty journalists, which had never happened until then. We built up strong relations with the press, with visible results.

The circumstances were far from ideal, and we could rely on no one but ourselves. I set about things in earnest, partly out of a sense of duty and partly to continue the work I had begun in the seventies. Many considered the post of EPP President to be a lesser job, something suitable for a former Prime Minister, but I found it an important task. To a certain extent I was returning to the position I had been in thirty years earlier: powerless, were it not for the power of my convictions.

It was not only the practical circumstances that left much to be desired. Politically speaking the idea of Europe was at a low ebb. Since Maastricht, supranational cooperation – and the ideal of European federalism – had been in a state of crisis. Europe had become strongly polarised between those against and those in favour. In fact, little had changed since the ratification of the Maastricht Treaty, except from the perspective of the euro.

I was fully aware of the problems of European integration; one of the conditions for an effective remedy was making a diagnosis that in no way avoided reality. As I was no longer a Prime Minister, I was able to speak about this more freely. The Community and its Member States had put on a very poor performance, especially as far as the civil war in the former Yugoslavia was concerned. Our credibility had reached a low point and I was challenged on this topic on numerous occasions. Wherever I spoke, the first question I was always asked was this: "What has Europe done for Yugoslavia?" European institutions were paralysed at the time by this internal and external malaise. They had little or nothing to say. Whenever and wherever I could, I tried to answer the criticism by providing analyses and alternatives. Within the EPP we had decided to go on the offensive. In this crisis we continued to stand behind a strong federal Union, as we had committed to doing at our congress in Dublin. Even afterwards, and despite the fact that the British Conservatives had joined our ranks, under my leadership the EPP continued to support the ideal of a federal Europe.

The Tories

Dealing with relations between the EPP and the British Conservative Party was one of the most delicate tasks I faced as head of the Party. To a certain extent these relations dated back to contacts between Christian Democrats on the continent and the Tories under Prime Minister Edward Heath, who had led the party from 1965 to 1975 and guided Britain into the EEC. At an international press conference in Rome on 21 January 1966, he announced that his party wished to join the European Union of Christian Democrats (EUCD), the precursor of the EPP. However, in late 1966 the EUCD decided that the door would remain closed to non-Christian Democratic parties. The Germans were in favour of welcoming them but lost out to the Italians, the Dutch and the Belgians. This was a mistake of truly historic proportions. The history of the Conservative Party and the EPP, and perhaps even of Great Britain, might have been very different if we had the foresight and the courage then to allow the Conservatives into the family of Christian Democrats.

With the arrival of Thatcher, structural cooperation with the Christian Democrats became even less obvious; because of her virulent Euroscepticism, the EPP was hardly in favour of seeking closer relations with the British Conservatives during the 1980s. This changed radically in the late 1980s and early 1990s. Following the departure of the Partido Popular, only the British and Danish Conservatives were left in the European Democratic Group (EDG) in the European Parliament. Their size was to their disadvantage, especially given the fact that Parliament was growing and, moreover, was gaining increasing legislative power. In order to form alliances, they began to look in the direction of the EPP Group, with whom they continued to have a good working relationship. Though the demand for structural cooperation was directed to the Group and not to the Party, the matter of the British Conservatives raised the question of our real political strategy and views, much more than the Partido Popular's membership did: Was the EPP an exclusive club of continental Christian Democrats, or was it open to non-Christian Democrats who accepted its main principles?

A Party Matter

Immediately after the 1989 European elections, EDG Group leader Sir Christopher Prout asked to join the EPP Group. Though the British Conservatives were not seeking membership in the Party, EPP Group chair Egon Klepsch immediately sent the request on to the Party, because of the delicate nature of Prout's request. The Group was not in a position to solve the matter on its own. The EPP, which was then led by Jacques Santer, decided that the time was not ripe for a rapprochement. Prout understood very well their "decision not to decide". As such, the problem did not lie in Parliament, where cooperation between the EPP and the EDG groups had been excellent, but rather in the individual member parties. In his attempt to convince others of the advantages of structural cooperation, Prout conducted a form of shuttle diplomacy, visiting all the national party headquarters.

As the new EPP leader, I was immediately faced with the matter as well. As Prime Minister, on 18 June 1990 I had already had a lunchtime meeting at my residence with Prout, the Danish Secretary-General of the EDG Group, Harold Rømer and Chris Patten, chair of the Conservative Party, later Governor of Hong Kong and member of the European Commission from 1999 to 2004. A practising Catholic, Patten was perfectly acquainted with our political thinking. In fact, Patten was in essence a Christian Democrat leading the Conservative Party!

A spate of rumours about the possible cooperation between the EPP and the British Conservatives broke out in late 1990. Following a meeting of the European Council in Rome that December, there were numerous speculations that a secret agreement had been struck between Helmut Kohl and John Major. Of course, there was nothing strange in this, given that Major had become the new leader of the Tories. In early 1991 he made it clear that he was in favour of a rapprochement with the EPP. In fact, this meant that he had totally rejected Thatcher's stance barely three years after her famous speech in Bruges!

Nonetheless, there was considerable debate during the EPP Summit at Val-Duchesse on 13 April 1991, a meeting that was mainly dedicated to the matter of the British Conservative Party. On the table lay two letters, one from Prout and one from Patten, dated 5 and 11 April respectively, both agreeing to the EPP programme. This was certainly considered significant, but was still not enough to remove suspicion and opposition, especially among Christian Democrats from the Benelux, the Democrazia Cristiana and Fine Gael. In the absence of those concerned, there was considerable debate on Christian Democratic identity, the constellation of party politics in Europe and the new developments within the Conservative Party, as well as the timing and modalities of possible cooperation. The answer to the question of membership would indeed be crucial for the future of Christian Democracy in Europe. At the same time, the need for expansion within the EPP was stressed. Seen from this perspective, Helmut Kohl was in favour of their joining and was supported in his motion by José María Aznar, Jacques Santer and Constantine Mitsotakis. Ruud Lubbers, the Italians and the Belgian Christian Democrats were totally against the motion.

Fraktionsgemeinschaft: Yes or No

A way out was offered to us by Klepsch, who put forward the idea of a community of groups or a *Fraktionsgemeinschaft*, analogous to that of the CDU/CSU group within the Bundestag. The compromise reached between those for and against an alliance was a combination of durability and difference. It was agreed from the outset that this new form of cooperation would only be temporary, that is, until the 1994 European elections. At that time, the situation would be evaluated and a decision would be taken as to whether the alliance would continue or not. Following that, on 8 May, the EPP Political Bureau agreed to the decisions taken at the EPP Summit of 13 April. As a result, eight working groups were set up with the purpose of examining congruence between EPP and EDG positions in various areas of policy. From an organisational point of view, it was proposed that in the future an extra EPP Group vice-chair and a deputy Secretary-General be appointed from the EDG Group.

In the meantime, during meetings of the Political Bureau, Klepsch reported on cooperation between the EPP and EDG groups. In general the conclusions were favourable but the Maastricht Summit cast a shadow over the proceedings. The EPP wanted to know Major's position before saying yes. Disappointment with regard to his performance in Maastricht was also thrown into the mix in reaching the final decision. The reports from the eight working committees were made available at the beginning of 1992. They presented no great problem as such but it must be said that the initial enthusiasm with which they were begun had largely evaporated.

This became clear during the difficult negotiations at the EPP Summit of 14 February 1992, which was also entirely dedicated to the topic of the British Conservatives. Lubbers succeeded in making a breakthrough on the day by offering an opening for a solution which stated that the Group could accept forms of cooperation and membership that were broader and more far-reaching politically than the Party was ready to accept. We already knew that the Tories would probably never agree to the Party programme. But we could indeed arrive at very fruitful cooperation within the European Parliament. Moreover, it was believed that the period of convergence should end and that a decision should be definitely reached. The deadline for such a decision had been set for the first quarter of 1992 but, because of the elections in Great Britain and Italy, it was postponed until 1 May. In fact, the leaders of the EPP were divided on the modalities of cooperation, but nonetheless remained committed to the idea of a community of groups or *Fraktionsgemeinschaft*.

The ball was once again in the EPP Group's court, but they dropped the idea of a *Fraktionsgemeinschaft*. Ultimately, individual associate membership (or "allied members") was opted for, which was also provided for in Article 5b of the Group's internal regulations. This formula was much simpler and would not result in the Group being divided into two factions. The Tories' acceptance of the solution constituted a victory for the EPP. Allied membership was passed by the Group in Strasbourg on 9 April 1992. Each individual MEP had to be voted on separately. For the Tories this also meant that they would agree to the decisions of the 1990 EPP Congress in Dublin. As a result, on 1 May 1992, thirty-two Britons and two Danes became members of the EPP Group. From then on, they had a right to speak on the Group's positions, could vote on and were also bound by its decisions. Given this form of adhesion, it was believed that the British Conservatives would move closer to the EPP, also as far as their position regarding the future of Europe was concerned. In those initial years of cooperation, I was not alone in believing that the Tories might become a member of the Party but, in the end, a certain amount of scepticism would not prove unfounded.

Nonetheless, cooperation did get off to a promising start. Under Prout's leadership, who as former chair of the EDG had now become vice-chair of the EPP Group, those in favour of European integration were dominant up until the 1994 European elections. During the EPP Summit in Brussels on 22 June 1994, the first after the European elections, there was doubt that cooperation would continue. Indeed, the British Conservatives did not apply for EPP Party membership in the same way that other Conservative parties did. Their cooperation with the EPP Group would become the cause of growing concern for me, and fifteen years later would result in a crisis with their new leader David Cameron. The Danish Conservative People's Party, Det Konservative Folkparti, on the other hand, opted for integration and became a full member party of the EPP in 1995.

Strong Foundations

As I was no longer Prime Minister of Belgium, from March 1992 on I could dedicate myself entirely to my task as leader of the EPP. I was now free to prepare more thoroughly for the following congress. My goal was to place the inspiration for our political action in the spotlight. This had not been done since the foundation of the EPP in 1976. I set to work straight away as I have always done.

I was surrounded by a team of young colleagues from the various political foundations and think tanks of the respective member parties, one of whom was Klaus Welle from Germany, who would later succeed Thomas Jansen as EPP Secretary-General in late 1994. I chaired the programme committee and, until the autumn of 1992, we worked regularly and thoroughly on reforming the Party and on developing a new Basic Programme. In fact, this had been on the EPP's agenda since the late 1980s, but it was only when I took over as President, and because of my availability, that things began to move and we were able to achieve results in a relatively short space of time. The membership of the Partido Popular, the alliance with the British Conservative Party and the attraction the EPP had for other non-Christian Democrat centre-right parties all posed considerable challenges in this respect.

Renewing the Basic Programme was for me a matter of fundamental importance and a high priority. I was convinced that the expansion of the Party would only prove durable and fruitful if there was agreement about the Party's political foundations. Moreover, the acceptance of these fundamental principles had to be a basic condition for membership for new political parties. It was a fact that the greater the difference in parties the more important the common basis became. This was our only true raison d'être, for one does not succeed through *Realpolitik* alone.

However, not everyone was convinced of the need for a new programme. Helmut Kohl, for instance, believed that paying such exclusive attention to the programme could prevent the necessary expansion of the EPP. I recall how I managed to convince him of the importance of renewing our programme and also of dedicating the congress in Athens to that topic. Reflection on our foundations was also crucial for the times we found ourselves in. Given the failure of Thatcherism and the confusion about Socialism in the West on the one hand, and the collapse of Communism in the East on the other – leading to a true *horror vacui*, in fact – the hour of Christian Democracy was on the verge of dawning.

The member parties from the Benelux had an important role to play in bringing about this new programme. Indeed, inspiration did not arrive overnight or from nowhere. The first draft was drawn up by the Dutchman, Jos van Gennip, director of the CDA's Research Institute. A rather clumsy translation of the texts initially caused some hilarity – unjustifiably so. In fact, the programme was redrafted by a few Belgians on that occasion. I was able to draw on the expertise of Flemish- and French-speaking Christian Democrats in the same way I had done during the founding of the EPP. The crux lay in reaching a consensus, something which we succeeded in doing marvellously.

The final draft was discussed in detail from 11 to 13 November 1992 at the EPP Congress in Athens, which was hosted by Prime Minister Constantine Mitsotakis. The original draft was amended on numerous occasions but without compromising its essence. It forms a document of reference *par excellence* for our political family to this very day, as I explain in the epilogue. The Basic Programme of Athens treats each pillar of the EPP in detail: a vision of society and humanity inspired by Christianity, a federal Europe and a social market economy. Moreover, the prologue focuses on changes within European society and the impact of these changes on the conduct of politics.

Hellenic Shore

The fact that the 1992 Congress was held in Athens stemmed from the strong anchoring of Nea Demokratia (ND) within the EPP. I met the late Konstantinos Karamanlis, the founder of Nea Demokratia, at my first meeting of European heads of state and government in 28 May 1979, also in Athens. We had gathered there to sign the document ratifying Greece's EEC membership. Karamanlis had the great foresight to orient his party towards Europe and lead his country into the Community, despite fierce opposition from Andreas Papandreou's Socialists. In fact, he is one of the founders of the European Union. A few years later, I developed good relations with ND leader Constantine Mitsotakis. He was Prime Minister during the re-launching of Europe in the early 1990s and also during the EPP Congress in Athens. His daughter, Dora Bakoyannis, was mayor of Athens during the Olympic Games in 2004 and at present is Minister for Foreign Affairs. Sadly, her husband, whose name she still bears, was murdered in a terrorist attack.

Ideas that two centuries later would become central to European thought can be found in the writings of the Greek revolutionary, Rigas Velestinlis-Feraios (1757–1798), an eminent representative of modern Hellenism. He was a man of the Enlightenment, a gifted politician, military tactician-soldier and grammarian, a rebel in the cause of Greek independence and a martyr, but most importantly a visionary.

He wrote the first constitutional charter for a democratic republic of the Balkans, including a declaration on human rights and a constitution that would allow the Greeks and all other peoples of the region to rule the republic democratically. This republic could only be consolidated through solidarity. He wrote, "The Bulgarian must be moved when the Greek suffers. And the Greek for the Bulgarian and both for the Albanian and for the Vlach" (*Human Rights*, Article 34). After the revolution no single people will be absolute masters, but all citizens will become sovereign without distinction of creed or language: "Greeks, Bulgarians, Albanians, Vlachs, Armenians, Turks and all other peoples" (Article 7 of the Constitution).

For Rigas, democracy was an ideal of a supranational order. In his new democracy all citizens would be equal before the law; religious services would be independent of state control; the language of each people respected; the lives and possessions of all protected. This is the vision we are now trying to realise in Europe. The figure and ideas of Rigas serve as an example for Greek schoolchildren, and also an example for European integration.

Rigas was also a martyr. He was betrayed and handed over by the Austrians to the Ottoman rulers in Belgrade. On the night of 24 June 1798, he was strangled, together with his companions, in a fort near the mouth of the river Sava and their bodies were thrown into the Danube.

Nea Demokratia embraced its EPP membership with open arms. As the Partido Popular had also recognised, the EPP offered a powerful European platform. Various EPP meetings were organised in Greece, which benefited the Party considerably. Since March 2004, ND is once again in power and Greece has a pro-European Prime Minister, Kostas Karamanlis, Konstantinos' nephew and two-term EPP Vice-President. Despite ND's non–Christian Democratic tradition, the party has demonstrated to the EPP the public importance of religions and the role of the Orthodox Church in particular. As a result, structural dialogue on monotheist religions has been endorsed at the heart of the EPP Group.

Nea Demokratia has also influenced the EPP's international outlook. It is thanks to a 2006 ND amendment in our statutes/internal regulations that we have today an EPP Secretary of External Relations. As well, Dora Bakoyannis has led with success the EPP Foreign Ministers' meeting, which was launched in 2007. In the troubled Balkan region, Kostas Karamanlis, in his capacity as EPP Vice-President, launched in 1999 the Western Balkan Democracy Initiative in an effort to identify and to propose potential partners for the EPP – today, most of these parties are observer or associate EPP members. It was also Kostas Karamanlis who, on behalf of the EPP, addressed in late 2003 the first congress of the AKP, the party of Turkish Prime Minister Recep Tayyip Erdoğan. In fact, Kostas Karamanlis is openly in favour of Turkish membership in the EU as long as it fulfils all the conditions of the accession process. Notably during a dinner with Giscard and me, he underlined his preference for dialogue with Ankara and not conflict. This position has become part of Greek government policy. It is definitely worth reflecting on the fact that Greece had the courage to adopt such a position, particularly given their age-old tensions in their relations.

When I look back and reflect on all these experiences I am deeply moved. They are summed up so beautifully in Ivoor en Brood (Ivory and Bread) from the collection of poems *Najaar van Hellas* (1947) by the Flemish poet, Anton van Wilderode

> O Hellenic shore where our fathers dwell
> The spacious halls of our wisdom's home
> Of your merit still all nations tell
> Where culture ground to crumbling stone
> Upon your hills we first were blessed
> Under your skies, Europe's spirit was born
> And all around your seas without rest
> like flutes to a feast, do murmur and call.

Unity in Diversity

About a year later, a new plan of action concerning policy options for the 1994–1999 legislature was passed at the EPP Congress, which was held for the first time at the site of the European Parliament in Brussels on 8–10 December 1993. A *rapporteur* was appointed for each chapter, including, among others, Thomas

Jansen, Klaus Welle, the Belgian MEP Fernand Herman and the Spanish MEP José María Gil-Robles junior.

In retrospect, the timing of the conference proved unfortunate. It was held much too soon as far as drawing up a campaign for the 1994 European elections was concerned. No one was really busy with the elections at that moment. Moreover, on 9–10 December there was a meeting of the European Council, which brought to an end the successful Belgian Presidency of the European Union for the second half of 1993. Though our heads of government were present at our congress all their attention was directed towards the European summit. That proved to be a serious miscalculation on our part. The press did not carry a single mention of our plan of action, which was titled "Europe 2000: Unity in Diversity".

"Unity in Diversity" – which later became the EU's motto for the treaty on a European Constitution – also became the motto of my presidency. It summed up the EPP's main ambitions: to become an open yet strong political family in the new European Union. Secretary-General Thomas Jansen gave me his complete support in the matter. And the interim results were appreciated for their worth. I was easily re-elected as President for another period of three years. This took place during the meeting of the Political Bureau on 19 May 1993.

Putting forward my candidacy for MEP in 1994 was a logical consequence of my work as EPP President. Both Leo Tindemans, who headed the electoral list, and I, in second place, received a considerable number of first-preference votes for the Flemish Christian Democrats. Tindemans was already an MEP and had been since 1989; I resigned from the Belgian senate to fulfil my new duties as an MEP. The high number of first-preference votes was a great encouragement, following my "journey through the desert" and my work "on the barricades".

Defeat at Corfu

During the campaign for the European elections a successor was also sought for Jacques Delors as President of the European Commission. The EPP played a prominent role in this for the first time. A number of candidates from our political family had put themselves forward for the position: Dutch Prime Minister Ruud Lubbers, Belgian Prime Minister Jean-Luc Dehaene, the Irish former Commissioner, Peter Sutherland, who had been proposed by his party leader John Bruton, and British European Commissioner Sir Leon Brittan, who as a Conservative Party member had been put forward by Prime Minister John Major. Only Lubbers and Dehaene were considered to have a chance of being elected. It was more or less certain that the new President would be from the Benelux and would be a Christian Democrat. The EPP could not let this exceptional chance slip by. There was, nonetheless, a real danger that neither candidate would survive the struggle. If we had put forward only one candidate, it would have seriously strengthened the EPP's position.

This is why I organised two attempts at negotiation. For the first, Lubbers, Dehaene and Prime Minister Jacques Santer of Luxembourg and I met in mid-June at the home of a notary public near Maastricht. Lubbers had arranged to meet there. This informal meeting was orderly but rather stiff. Lubbers apologised to Dehaene for his remark about Dehaene not being capable of the job. Dehaene rejected Lubbers' proposal that they both pull out in favour of a third candidate. All I could do then was to restore contacts and accept the political impasse.

The second confrontation took place during the EPP Summit in Brussels on 22 June 1994. Surrounded by TV cameras, a relaxed Dehaene sat down next to a smirking Lubbers, both under Kohl's critical eye. Press interest was enormous. This personal duel, a match between Belgium and Holland, was a feeding frenzy for the media. Both protagonists were photographed for minutes on end, much to Kohl's annoyance, as he could already see the storm clouds hanging over the EPP, precisely because the two candidates were parading in front of each other in public.

Each candidate introduced himself before the discussion started. Lubbers stressed that he had been Prime Minister of the Netherlands for the past twelve years and, as such, was also a member of the European Council. He also pointed to his good relations with the smaller countries and the candidate member countries, and emphasised the role of "big little state" played by the Netherlands in the EU, its complementary role in the French–German axis, the support he had received from various countries for his candidacy and his enormous popularity at home. He added in closing that he appreciated Dehaene's qualities and was willing to accept a consensus.

Dehaene, on the other hand, warned of a conflict between the European Council and the Parliament if the Council should not reach a decision. He understood that it was impossible to reach a unanimous decision during the EPP Summit. He called for calm within the EPP. According to him, his candidacy was supported by other members of the European Council, who had been positive about Belgian EU Presidency. He was available and willing to continue his work in Europe by becoming President of the Commission.

Kohl then announced his preference for Dehaene. He had made up his mind and had already reached an agreement on the matter with François Mitterrand some time ago. However, they had failed to consult John Major on the subject, which was to prove fatal to Dehaene's candidacy. On previous occasions, they had discussed Delors' appointment and re-appointment with Thatcher. In addition to this, a conflict was brooding between Kohl and Lubbers. Much to Kohl's annoyance, Aznar, who was still an opposition leader at the time, spoke out in favour of Lubbers. This did little to improve matters between them. At no time would Kohl ever accept a compromise. I could do little else than reach the same conclusion: we had two good candidates. The whole thing was a very embarrassing experience for the EPP, because on June 1994 the European Council in Corfu was stalemated on the nomination of the new Commission President. Council President Andreas Papandreu lacked the capacity to reach a consensus.

Finally, a Christian Democrat from the Benelux was elected as President of the European Commission. Helmut Kohl put forward Jacques Santer as the new President during a special meeting of the European Council on 15 July. Given his many years of experience as a Prime Minister, the choice was obvious. He had also been President of the EPP for four years and shared our convictions on Europe. We placed much hope in his presidency, which, unfortunately, would not end favourably. Because of events beyond his control he was judged negatively in retrospect, which was unfair.

Chapter VII

The Fall of Santer

The five-year period during which I was EPP President and also chair of the EPP Group in the European Parliament demanded considerable energy. I tried time and again, with the extremely limited means that were given me, to reach consensus in a Parliament that I did not as yet know very well. Not a day went by without my being preoccupied by it. A completely new direction had been taken in electing me, a complete newcomer, as head of the Group. I could indeed draw on twelve years of experience as Prime Minister and member of the European Council, and on the fact that I had been EPP President since 1990. My ambition was to lead the Group with authority and to ensure a greater degree of cohesion in its actions. To succeed, I would have to overcome much opposition and make many personal sacrifices.

There are two crucial matters that I shall never forget. My presidency coincided with Jacques Santer's leadership of the European Commission. The vote of confidence for Santer as President and his subsequent fall from grace cost me a great deal, personally and politically. I had fought hard to defend his Commission both within our Group and in Parliament. At times it felt as if I were tilting against windmills. Besides this particular challenge, Forza Italia's joining the EPP Group proved to be one of my most delicate tasks as Group leader. The case was extremely complex, not to say explosive, both in Italy and in Parliament as well as on my home front. My own Flemish Christian Democrats would punish me for my support for Forza Italia's membership.

The role played by the chairs of political groups is of vital importance for the European Parliament and for European political parties. Many get the call but very few are cut out for the task. From day one, one is never fully the boss of a group of such a variety of politicians – different in nationality, sensitivity and competency. Leading a group is no easy task and authority is something one has to earn. Also, the atmosphere at the European Parliament is totally different from that of national parliaments. Other customs and procedures are at work, and one is much further removed from the electorate compared to national politicians.

The Parliament has been granted legislative powers since the Maastricht Treaty; as a result the role of a group chair has become more important. But we also had the task of increasing our power base. Who could we best join forces with in order to carry through our policy preferences? In fact, what this challenge amounted to was increasing our power in Parliament to the detriment of the Socialists. This was of fundamental importance, particularly for the Germans. The leading figures within the EPP were convinced of this and believed that I could quickly bring it about. What we had fought so hard for in Maastricht now had to be put into practice. The combination of Party and Group leadership – very much to Helmut Kohl's liking and very much in keeping with the German tradition – was supposed to help in this respect. During that five-year period I was able to increase the Group's membership by a third, from 157 to 201 MEPs.

My Election as EPP Group Chair

Our group had had three leaders since the first direct elections to the European Parliament in 1979. Egon Klepsch from Germany was Group leader in 1977 and remained so until he became President of the Parliament in 1992, with the exception of the two-year period in 1982–1984, when the Italian Paolo Barbi took on the post. Leo Tindemans succeeded Klepsch in 1992 and remained chair until the elections in June 1994. In keeping with group tradition, the chair is elected before the first plenary session of Parliament. Also in keeping with tradition, the five-year legislative term is divided: the President of the Parliament is elected at the beginning for a period of two-and-a-half-years, followed by the election of a new President for the second two-and-a-half-year period. This is also the case for all posts such as chairs and vice-chairs of committees, delegations, coordinators and so on. In fact, this tradition has had a negative effect on the influence and visibility of Parliament.

Klepsch and Tindemans did not get on very well. My cooperation with Tindemans during his period as chair went very well. It was a matter of course that Party and Group leaders should cooperate closely, to ensure harmony and maximum effectiveness for the EPP as a whole; this is also why the Group chair is an *ex officio* Vice-President of the Party. I also got on well with Klepsch, despite discontent over his lack of decisiveness – in informing the Party of the British Conservatives' request for membership, for example – and about the way he led the Group. The Presidency of the Group, the daily business meetings comprising the chair and the vice-chairs, was more like the chair's personal court. Helmut Kohl had his own plans for the Group and for Parliament, and they were quite radical. He had supported Klepsch for years but considered Klepsch's political career at an end following his presidency of the Parliament in 1994. Nor were Tindemans and Kohl the best of friends. Kohl had said in 1992 that he did not agree with Tindemans being chair. When he was elected anyway, this displeased Kohl considerably and he used his influence when the first opportunity arose.

Kohl called a meeting at the *Kanzlei* in Bonn in 1994 for the election of a new Group leader. Striking by his absence was Günther Rinsche, the leader of the German delegation who was a good friend of Kohl's. Rinsche had in fact supported Tindemans during his election in 1992. Kohl's message was clear: Tindemans should not be chair but I should. As soon as I was no longer in the running for President of the Commission it was as clear as daylight to Kohl that I should play a full role in Parliament. Tindemans was extremely disappointed and withdrew his candidacy. I was elected leader by a clear majority, with 118 of the 136 votes cast.

At the June elections in 1994, the EPP Group won 157 seats (compared to the Socialists' 198 and the Liberals' 43). The Group comprised the delegations of the EPP member parties and British Conservative MEPs who were allied with the Group but were not members of the Party. We had won 40 seats more than in the previous European elections, when the British Conservatives were not yet allied with our group. We had achieved a status quo in comparison to the end of the previous legislature, as had the Socialists and the Liberals. So the balance of power in Parliament did not change drastically. Things were quite different within our group. Our Italian delegation had shrunk as a result of the implosion of the Democrazia Cristiana. Moreover, the Conservatives had suffered a heavy defeat in the elections in the UK. Fortunately, these setbacks were compensated for by the good results achieved by the CDU/CSU and the Partido Popular. The latter managed to almost double the number of its MEPs.

The Santer Commission

The Prime Minister of Luxembourg, Jacques Santer, was put forward on 15 July 1994 by the European Council to be the new President of the Commission. The Parliament would first debate this proposal and voice its opinion through an actual vote of confidence on his candidacy. There was considerable opposition to Santer, both among those in favour of further European integration and those against it, and also to the way Jacques Delors' succession had been handled. Moreover, this also proved to be a difficult matter for me personally. The issue was handed to me as the new Group chair, and I do not know what my political destiny would have looked like had I failed to steer Santer safely through Parliament. It would have been a disgrace for an EPP President and Group leader to fail to gain a majority in favour of installing a Commission President belonging to the EPP, who had, moreover, been proposed by a European Council that included many EPP government leaders.

Before the vote, the Conference of Presidents – the body of chairs of the different groups in the European Parliament – was held, during which Kohl defended Santer's candidacy. After this, Santer spoke to each political group separately. The mood turned against him very quickly, despite the fact that he had been proposed unanimously by the European Council. The resistance came mainly from the Socialist side, who ignored the fact that their respective Socialist government leaders in the Council had spoken in favour of him. It was not surprising that people were watching how the Socialists would vote; they formed the largest group in the new Parliament. The animosity over Corfu was still very much alive. Through this vote, many MEPs were out for revenge and were convinced that Parliament could make its influence felt through it.

The closer we got to the vote the stronger the opposition seemed to become. I remember listening to the news coverage on the morning of the vote, and all the commentators were convinced that Santer would not make it. Things became so dramatic that Kohl phoned me up just before the plenary session. "You will fight", he said forcefully, and immediately got in touch with all his fellow heads of government. Following Kohl's move, Felipe Gonzalez managed to convince the Spanish Socialists to vote for Santer. But not all heads of government remained faithful to their commitment to Santer.

Because there was so much at stake, I suggested to the EPP Group members that we hold a party-line vote. Tindemans ignored the Group's decision and abstained from voting. Santer was really furious about this. In the end, Santer won by a narrow margin: 260 in favour to 238 against with 23 abstentions. We can thank the support of the Spanish Socialists, inspired by Felipe Gonzalez, and the MEPs of the Partido Social Democrata (PSD), the party of the Portuguese Prime Minister Anibal Cavaco Silva, who was then a member of the Liberal Group, for tipping the scales in favour of Santer. The new Santer Commission entered office only on 23 January 1995, having once again gained a narrow majority in a vote of confidence on 18 January.

The fact that there was extensive debate in Parliament about the new Commission President was a direct result of the Maastricht Treaty. For the first time, special hearings were held in the parliamentary committees to sound out possible candidates for the office of Commissioner. That did not go very smoothly. In fact, this involved a debate on confidence as well as hearings in the committees, a process that was later adopted in Treaty articles.

In the Santer Commission, seven of the fifteen Commissioners were part of our political family. There were some unexpected successes like Mario Monti, Commissioner for the Internal Market, put forward by Italian Prime Minister Silvio Berlusconi, the leader of Forza Italia. Monti was very close to the EPP and so, by accident, as it were, we could rely on this excellent Commissioner. The Dutchman Hans van den Broek and the Englishman Sir Leon Brittan remained on from the previous Delors Commission. The Spaniard Marcelino Oreja was new but a stalwart. Following the accession of Austria to the EU in 1995, the government delegated Franz Fischler, who would hold the post of Commissioner for Agriculture for ten years.

"Neutral" Enlargements

During the 1994-1999 legislative term, relations with the British Conservatives were not always easy but were positive on the whole. They too appreciated the cooperation with the EPP Group. With considerable effort, I managed to have common positions adopted within the Group, including by the British. Sometimes they still voted in accordance with their own voting lists, but differences of opinion were aired internally. The quest for a consensus was essential so that we could continue to work towards areas of legitimate cooperation.

Happily, not every expansion of the EPP was controversial. The Österreichische Volkspartei (ÖVP, Austrian People's Party) was an obvious candidate for membership because there had already been a long tradition of cooperation with them. But in the wake of our alliance with the Tories, the EPP received requests from other Conservative parties that also had no historical links with Christian Democracy. The Nordic Conservative parties, for example, were not Thatcherite; they were pro-European and often progressive, socially speaking.

Contacts with new member parties were prepared in detail and well beforehand. Secretary-General Thomas Jansen and I went on a trip to Scandinavia for this purpose in 1992 in the run-up to the enlargement of the EU. In Sweden, for example, we met pro-European Prime Minister Carl Bildt and Alf Svensson, the equally pro-European Vice-Premier of Bildt's centre-right coalition government (1991-1994). Carl Bildt was the leader of the Moderata samlingspartiet and Alf Svensson the leader of the Kristdemokratiska samhällspartiet, known since 1996 as Kristdemokraterna.

In cooperation with Svensson I worked out the so-called Mellanfjärden formula, after the eponymous idyllic village where we set out the steps for their accession to EPP membership, in keeping with the Party's statutes. These parties would first become observer members, then associate members if their country became a candidate Member State, and finally full members, as was the case for the Nordic countries in 1995. Two prominent female politicians of Moderata – Margaretha af Ugglas and Gunilla Carlsson – served as Vice-Presidents of EPP and continued to be active in our work.

Finland also had a small Christian Democratic party – Suomen Kristillinen, known as the Kristillisdemokraatit since 2001 – but because of their Euroscepticism, they were complicated negotiating partners for us. The conservative Kansallinen Kokoomus (National Coalition Party), on the other hand, was very cooperative and had a strong European orientation; they played a very active role in the EPP. The then President of Kokoomus, Sauli Niininstö, was the central figure during the negotiations with the EPP. He was also the last chair of the European Democrat Union (EDU, a looser Europe-wide centre-right organisation) and one of the chief architects of the EDU's integration into the EPP; as recognition for this achievement, he was elected Honorary President of the EPP. He was succeeded as party leader by Ville Itälä, who became an MEP in 2004. Since then, the party has been led by the young Jyrki Katainen. In 2006 he was elected Vice-President of the EPP; in March 2007, his party did very well in the general elections in Finland, gaining an extra 10 seats to bring its total to 50, only one less than Prime Minister Matti Vanhanen's Centre Party. Katainen was appointed Deputy Prime Minister and Minister of Finance. In this latter capacity, he also leads an informal gathering of EPP Ecofin Ministers.

Also, soon after the Danish Conservative MEPs joined the EPP Group, the party itself opted for full accession to EPP membership. Their party leader and long-standing Prime Minister, Paul Schlüter, played a key role in bringing about the Single European Act and the Maastricht Treaty. Interestingly, the Brussels-based Danish Conservative, Harold Rømer, the outgoing Secretary-General of the European Parliament, and Niels Pedersen, the former Secretary-General of our group, held these important posts despite the relatively light "weight" of their party compared to other EPP member parties. Unfortunately, the DKF is no longer the leading party at home. Since 2001, led by former party leader and Deputy Prime Minister Bendt Bendtsen, who was followed (since September 2008) by Lene Espersen, they have been junior partners in a coalition with the Liberals.

Unfortunately, the Norwegians rejected membership in the EU on two occasions, the 1972 and 1994 referenda. This did not prevent us from having good relations with Høyre, the Conservative Party of Norway, the most pro-European party in the country. Former party leader Jan Petersen was a very successful Foreign Affairs Minister in Kjell Magne Bondevik's second government. Petersen was also a key figure in bringing about the EDU's integration with the EPP, and he also energetically led the EPP's Foreign and Security Policy Working Group. Bondevik, a minister in the Lutheran church of Norway, is the longest-serving non-Socialist Prime Minister of Norway; he was in office from 1997 to 2000 and from 2001 to 2005. The EPP went to considerable lengths to bring Høyre and Kristelig Folkeparti closer together, thus making the last Bondevik government possible. The Kristelig Folkeparti (KrF) was smaller than Høyre but, because of Bondevik's inspiring leadership, the party could not be ignored politically speaking. But the KrF's anti-EU stance has limited its involvement in the EPP to being an observer member. Høyre, on the other hand, an active EPP associate member since 1995, campaigned vigorously for EU membership prior to the two failed referenda and remains committed to a future attempt.

The Nordic member parties and the Moderaterna in particular have had considerable influence in formulating aspects of economic policy in our programme. They have much in common with the British model, whereas most of the other EPP member parties traditionally follow the so-called Rhineland model. When the Nordics fail to carry the vote, they always follow the majority decision. They also play an important role in promoting innovations to the EPP's political leadership. Finnish Deputy Prime Minister Jyrki Katainen and Swedish Prime Minister Fredrik Reinfeldt are fine examples. Reinfeldt was an active leader in the EPP youth movement. He belongs to the first generation of our political family to effectively come of age within a European political context. He radiates trust in the pro-European stance of the EPP and the positions of our Conservative member parties from Nordic countries.

Unfortunately, this does not hold true for the Christian Democratic parties in the Nordic countries. With the exception of the Swedish Kristdemokraterna, these small Protestant Christian parties are invariably Eurosceptic; they differ strongly in this respect from their Nordic Conservative counterparts. At the same time, these parties maintain very conservative stances on ethical issues, as a result of which they are often unfairly branded as reactionary. In my opinion, they lack one of our basic principles, which is a commitment to Europe and to our belief that integration should lead to a federal Europe.

Hostage in Porto

The fact that a pro-European stance is of vital importance to a decision whether to award EPP membership was abundantly clear in the expulsion of the Portuguese Centro Democrático Social (CDS, Democratic and Social Centre) in 1993. This party had become increasingly nationalist and opposed the ratification of the Maastricht Treaty. Even the British Conservatives had ratified the Treaty, albeit with certain opt-out powers.

Contacts with Portuguese member parties date back to the days following the Carnation Revolution. In early summer 1974, I and my Belgian colleague, Charles-Ferdinand Nothomb, President of the French-speaking Christian Democrats, met the founder of the Partido Social Democrata (PSD), Francisco de Sá Carneiro. This brilliant political leader, who had emerged during the Carnation Revolution, was travelling through Europe in search of political partners. We had frank and hopeful talks but a few weeks later we heard that he wanted to join the Socialist International. He was refused membership and ultimately chose to align himself with the Liberals. His contacts with Hans-Dietrich Genscher, the leader of the German Liberals, most likely had something to do with it.

Sá Carneiro became Prime Minister of Portugal in 1980, forming and leading a coalition with the CDS. Unfortunately, I never met him again as he died in a tragic accident on 4 December 1980: both he and his life-long partner (*sua companheira*), Snu Abecassis, and the Minister of Defence, the Christian Democrat Adelino Amaro da Costa, lost their lives when their plane crashed on the outskirts of Lisbon. They were on their way to a meeting in Porto during the campaign for the presidential elections. Many Portuguese are still convinced that there was sabotage involved. When I attended the funeral ceremony for the two ministers in the chapel of the Jeronimos Abbey in Lisbon, I could not help thinking back on the attack on the Crystal Palace in Porto. I had experienced those events there together with CDS President Diogo Freitas do Amaral and his closest colleague, Amaro da Costa.

Because Sá Carneiro had joined the Liberals, Freitas do Amaral's Centro Democrático Social became an option for us. In autumn 1974, I took part in the foundation of the CDS at the Crystal Palace in Porto as a member of a delegation of Christian Democrats from Belgium and the Netherlands, along with other like-minded delegates from all over Europe and from President Giscard's youth movement. There from the Netherlands was Roelof Kruisinga, very active at the time and later Minister of Defence. He resigned from the government, however, in protest against American plans for a neutron bomb.

That afternoon, the Crystal Palace was attacked by communist militants. They blocked the main entrance and locked the participants inside. The police looked on powerlessly. We were kept hostage all night and many feared that the militants would invade the building. Kruisinga said manfully that we should be prepared to die for our ideals. But the siege came to an end following a phone call made by one of Giscard's youth. They were able to alert the French President to our perilous situation. He urged the Portuguese President to take action. The President sent paratroopers to the Crystal Palace and we were freed by early morning. We had to get into cars that were still there in the car park. We sat there for hours – my companion in misfortune, Frank Swaelen, who was Secretary-General of my home party, and I – in Amaro da Costa's little car until the paratroopers had cleared the way and escorted us back to our hotels in the centre of Lisbon.

The CDS developed into a full-fledged Christian Democratic party and was in government on several occasions. Next to Freitas do Amaral, Francisco Lucas Pires was one of the prominent figures and a leading member of Parliament. He and a few like-minded colleagues left the CDS following pressure from party president Manuel Monteiro, who had taken an anti-European line and opposed the ratification of the Maastricht Treaty. This is why the CDS membership in the EPP was revoked.

The PSD, on the other hand, remains a leading party in the country. I have maintained excellent relations with its former leader, former Prime Minister and current President of Portugal, Anibal Cavaco Silva. We were colleagues in the

European Council. Already then a rapprochement had developed between the PSD and the EPP. In 1992, Thomas Jansen and I visited Lisbon and held detailed talks with Cavaco Silva. He was in favour of joining the EPP and was prepared to leave the European Liberals, following the necessary preparations by both sides. Marcelo Rebelo de Sousa, who was party president from 1996 to 1999, became heavily involved and committed himself to this move; he assumed his responsibilities in this respect.

It was no coincidence that following the elections in 1994 the EPP Group held its workshops in Lisbon, where Minister for Foreign Affairs José Manuel Durão Barroso addressed the Group. In May 1996 a decisive meeting also took place in Lisbon between the leaders of the PSD and a delegation of the EPP comprising Helmut Kohl, José María Aznar and I. That same year, the PSD MEPs joined the EPP Group and later the PSD became a full member of the EPP. With this new member, our political family once again had a strong and trustworthy partner in Portugal.

One tangible result of the admission of the PSD to the EPP was the appointment of Mario David as Secretary-General of the EPP Group in 1997. He remained in office until 1999. Before that, he had been Secretary-General of the Liberal Group in the European Parliament. His appointment was a gesture of appreciation for the move made by the PSD MEPs, in which he played an important role. David was later elected Vice-President of the EPP at the party congress in Rome in late March 2006. In 1999, when Barroso was President of the PSD, he was also elected EPP Vice-President at the party congress in Brussels and, soon after, became Prime Minister. I will go into this in more detail later, but Barroso's election as President of the Commission in 2004 was a direct result of the EPP's intervention. My friendship with Cavaco Silva also continues to this day, and I was very touched when I was included as one of his guests of honour to the ceremony at which he was sworn in as President of Portugal in 2006.

A Role for the Parliament and Another for the EPP

In the meantime, things continued to move ahead in the EU. The European Parliament contributed further to the integration process. Five years after the signing of the Maastricht Treaty, an Intergovernmental Conference of government leaders was held once again. The purpose was to adapt European institutions to fit the enlargement of the EU. The structures of the EU dated back to the Europe of the six, which had since grown to fifteen and was preparing to include a large number of candidate Member States from Central Europe.

The IGC was held in Amsterdam on 16–17 June 1997 and led to a new Treaty – once again under Dutch Presidency. Amsterdam was not of the same calibre as Maastricht, however. It soon became apparent that the EU was insufficiently prepared for expansion but was in fact ready for deepening, that is, with regard to the powers and legislative role of Parliament. Some degree of success could be achieved as a result: the expansion of co-decision and giving a greater degree of control to the Commission, for example. Though the Amsterdam Treaty only took effect on 1 May 1999, following a second referendum in Ireland in 1998, it would be the cause of a great deal of disturbance in the European household.

In contrast to my roles in the Single European Act and the Maastricht Treaty, I was no longer directly involved in the Amsterdam negotiations. I did follow developments very closely, however. As EPP Group chair, I was often in a position to set out my point of view at plenary sessions of Parliament. In addition to being part of the daily business of the EPP, the new treaty was debated at a congress dedicated to that purpose, in Madrid on 5–7 November 1995. Hans-Gert Pöttering, vice-chair of the Group at the time, drafted a congress document on this topic. Moreover, as EPP President, I led the EPP summits during which preparations were made for European Councils, particularly on such issues as institutional developments, the enlargement of the EU and the gradual changeover to the euro. It was mainly this last topic that gave me hope, as it was one of the achievements of Maastricht to which I had contributed and which had a clear influence on our daily lives. Through the euro, Europe became something tangible in the hands of its citizens.

Amsterdam placed employment at the top of the political agenda for the first time. The European Social Chapter, which the British had opted out of signing in Maastricht, was integrated into the new Treaty. Though employment remained very much the competence of the Member States, our Toulouse Congress on 9–11 November 1997 was dedicated to this topic. A few days later, a special meeting of the European Council in Luxembourg was devoted to defining policy with regard to the high levels of unemployment. The social market economy is one of our basic founding principles but had hardly been developed till then. What we tried to work out at the time was a precursor of the Lisbon Agenda. But enlargement was on the agenda too, because decisions had to be made at the beginning of December.

The congress document on globalisation, unemployment and the expansion of the EU required considerable effort to draft. Given the various nationalist-inspired differences in policy it was far from easy to arrive at a consensus. The proposals put forward by the Nordics were at odds with Christian Democratic principles stemming from the Rhineland model. Moreover, a number of representatives from EU candidate countries had questions about the pace of European enlargement. Ultimately, an agreement was reached on a congress resolution drawn up by Hans-Gert Pöttering as well as on a concluding text entitled "We are all part of one world". Rather than agreeing to sweep certain differences under the carpet, the EPP showed once again that its strength lay in its capacity to conduct in-depth debate with a view to reaching a common programme.

Our Italian Mainstay

The Democrazia Cristiana had been one of the pillars of the EPP for years. The Italian Christian Democrats were founding members of our party and formed one of the strongest delegations in our group. Their leaders by tradition played a prominent role in the integration process. I also got to know how the party worked on the home front. I recall long congresses during which the various political streams battled for dominance. I knew many of their leaders personally and was present at Aldo Moro's funeral. Those were the years when the Democrazia Cristiana was omnipresent.

This all came to an end in the early 1990s. The Italian Christian Democrats lost electoral and political power, first at the national level and then at the European level. The Tangentopoli corruption scandal put an end to a party political regime mired in internal dispute, immobility and unstable government. Following successive defeats at the polls and the arrest of their leaders, one by one, the Christian Democrats seemed be in their death throes. A number of their representatives abandoned the sinking ship just in time to join other parties or set up new political movements.

Nonetheless, attempts were made to stop the disintegration and to re-group the party. Together with Helmut Kohl, I was invited to a congress in Rome on 27–29 July 1994, at which Rocco Buttiglione was elected Secretary-General – a position comparable to that of party president. During dinner that evening Buttiglione, who is more of an academic than a politician, asked me what one needed to be a good party leader. My answer was this: "Once you accept the job, you have to be available for your party twenty four hours a day. A good party leader should be a person of consensus, someone who rises above the various *correnti*, someone who is constantly aware of and informed about the political situation and is involved in the formation of his party representatives and concerned about the running of the party". Buttiglione could not produce a consensus and the Italian Christian Democrats ultimately fell apart under his leadership.

The partial introduction of the majority electoral system also proved detrimental for a party like Democrazia Cristiana that had occupied the centre for decades. A split occurred between centre-left and centre-right. As a direct successor of the Democrazia Cristiana, the Partito Populare Italiano (PPI) was founded in early 1994, the name referring to the pre-war Italian People's Party of Don Luigi Sturzo. The PPI increasingly attracted those with left-wing leanings, whereas the Centro Cristiano Democratico (CCD) attracted those from the right.

Initially, the EPP refused to take sides and supported every effort to reform the old Christian Democratic Party. Paradoxically enough, their relations with the EPP drove the two parties even further apart. Because the situation was escalating, and following threats of litigation, among other things, the EPP forced both parties to reach an agreement or else face the risk of losing their membership. All this took place on the margins of the EPP Summit at Cannes on 25 June 1995. In European terms, the EPP wished to provide a tent for all Christian Democrats from Italy.

All of this typified the deadlock: instead of worrying about the voters, they waged war for symbolic and personal power. The situation deteriorated dramatically and we felt the repercussions of it in Europe too. The 1994 election results were a catastrophe: the Italian Christian Democrats dropped from 26 to 11 seats. After the elections, it was suggested that members of Forza Italia be allowed to join the EPP Group. It was understood – and rightly so – that the majority of Christian Democratic voters had voted for Silvio Berlusconi's new political movement. Similarly, Berlusconi did not hesitate to point out that his movement was clearly rooted in the tradition of Christian Democracy in Italy.

However, the river that separated both parties was still far too deep to cross. It was mainly the controversial image of Berlusconi that formed an obstacle. There was no basis of support for achieving a quick rapprochement within the EPP Group, not in the least in Italy. No one was in favour of suddenly ditching the Italian Christian Democrats, with whom we had cooperated so well in Europe, for a political formation that not only raised various doubts in political circles, the media and public opinion but, moreover, had only existed for several months. Personally speaking, I was not in favour of immediately approaching Forza Italia and my position remained unchanged for years. Kohl's position must be understood in the same light: immediately after the European elections, during the EPP Summit on 22 June 1994, he spoke out against Forza Italia's membership in the EPP Group. Neither could we ignore the fact that Berlusconi had formed a coalition with the so-called separatists and post-fascists. Only Aznar was willing from the beginning to accept Berlusconi and Forza Italia into the EPP.

That Forza Italia explicitly refused to call itself a party created more doubt about Berlusconi and his intentions. Nonetheless, he did seek contact with the EPP from the beginning. I met him for the first time in July 1994. Given the sensitivity of the matter, discussions had to be extremely circumspect. I flew to Italy alone to meet him in private at his house in Milan. The conversation proved constructive and prospects of Forza Italia joining the family arose during the discussion, but things never went any further for the time being.

The Italian political party system continued to remain extremely unstable, the centre in particular being highly volatile. Because the Christian Democrats were divided into two camps – those who supported the government and those in opposition – the tendency towards regrouping to the left and to the right continued, albeit with difficulty and following various formations and flags. The PPI formed an alliance with Romano Prodi's Democratici for the snap general elections in April 1996. The Christian Democrats of the CCD formed an alliance with Forza Italia. Others went to the polls independently, like the Cristiano Democratici Uniti, for example. Excellent cooperation had been going on with Pierluigi Castagnetti's PPI delegation in Parliament. He belonged to that good tradition among Italian

Christian Democrats of trust in a commitment to Europe. But because the hope that the PPI would become a stable and strong centre party failed to materialise, discontent began to grow – particularly in Germany – at the idea that only left-wing Christian Democrats had found shelter in the EPP.

Initially Forza Italia's MEPs formed their own group, Forza Europa. At the time, internal regulations on the formation of groups in Parliament still permitted this on the basis of one (relatively large) party. Already in 1995, Forza Europa formed an alliance with Rassemblement pour la République (RPR), the French neo-Gaullists, and with the Irish Fianna Fail. Together they called themselves the Union for Europe Group and described themselves as right wing, nationalist and Euro-critical. Because a few Christian Democrats belonged to those elected for Forza Italia, there were regular contacts within the European Parliament. Forza Italia also continued to insist on closer cooperation with the EPP Group. Attempts were made to harmonise our actions in Parliament, particularly on policy issues. From 1997 on, formal meetings were held each month between Forza's head of delegation, Claudio Azzolini, the vice-chair of the EPP Group, Hans-Gert Pöttering, Gerardo Galeote, an MEP from Spain and Pierluigi Castagnetti, the PPI head of delegation. But this went nowhere in terms of formal cooperation.

Forza Italia After All

The situation changed dramatically in late 1997. Just before one of the plenary sessions in Strasbourg, I read by chance in a local newspaper that Forza Italia and the neo-Gaullists were planning to form a new European party. I was dumbstruck, as was Pöttering, who was sitting next to me. The founding of a new party, a durable alliance to our right, formed a serious threat to the future existence of the EPP, as its attraction for existing and future member parties of the EPP was not to be underestimated. Moreover, the EPP's efforts to include within it non-Christian Democratic yet like-minded parties would be completely undone. More than anyone else, Kohl feared a scenario in which the EPP would be split into a left and a right wing. You could have the best programme, but if you did not have the numbers there was little you could do, he believed.

That these parties were serious became clear at the press conference given by Berlusconi and the RPR President, Philippe Séguin, on 18 December 1997, at which they announced the foundation of their new party, Union for Europe. It was clear as daylight that the EPP had to undertake a counter-offensive. In the first place, we had to establish a permanent relationship with Forza Italia. An alliance with the neo-Gaullists would also considerably strengthen our position in France. I personally contacted the Frenchman, Jean-Claude Pasty, the chair of the Union for Europe Group in the European Parliament. He was certainly well-intentioned and I was able to develop an efficient mode of cooperation with him in Parliament.

But he also held on to the illusion that he could keep the group going even after the 1999 European elections. He was completely mistaken, since the neo-Gaullists joined the EPP Group and he did not return to the European Parliament.

The "Cottage" Talks on the Rhine

When the establishment of a new party was announced formally in the spring of 1998, things began to move very rapidly. Kohl took the initiative and on 24 March 1998 he invited all the important players to his cottage on the Rhine: Prime Ministers José María Aznar of Spain, Jean-Luc Dehaene of Belgium and Jean-Claude Juncker of Luxembourg, former Swedish Prime Minister Carl Bildt, German Minister for Finance and CSU President Theo Waigel, EPP Vice-President and CDU leader for Schleswig-Holstein Ottfried Hennig and myself. Klaus Welle, who was still Secretary-General of the EPP, had drafted a memo beforehand that was discussed in detail and amended during the meeting. I formulated a new text verbally, mainly because Welle could not be directly present at the higher levels of negotiation. Afterwards, I decided to put the decisions made at our meeting on paper and then ask for agreement from all those present. The goal was to set out a strategy that would strengthen the EPP structurally. It was, therefore, necessary that Forza Italia and the neo-Gaullists be allowed to join our family. A merger also had to take place between the two organisations that addressed the centre-right in Europe: the mainly EU-centric EPP and the looser, nationally oriented EDU. With a view to making the organisation more effective, we also needed to increase the number of EPP leaders participating in our summits.

Following years of fruitless discussions, the turns taken and openings made at this meeting permitted me as EPP President to allow the Party to grow to its natural maturity. The political vacuum created by the implosion of the Democrazia Christiana needed to be filled and there was really only one candidate that could so: Forza Italia. With the drive of RPR party leader Alain Juppé and with the support of President Jacques Chirac, the neo-Gaullists in France managed to shake off their Euroscepticism and were on the verge of approaching the EPP. Returning that night to Brussels by car I felt I was a happy man, but little did I know that vengeful gods were waiting in the wings to cast me from this height onto the rocks below. A few of the participants who found it difficult to let go of the EDU were less pleased. After the clear agreements made with Kohl, it was inevitable that the organisation would be integrated into the EPP. This would take a couple of years, however, as personal interests and the sheer inertia of those in established posts served to slow down the process of transformation. But eventually, these interests proved unable to stop the forces of unity; the merging of the EDU into the EPP was finally and formally completed at the 2002 EPP Congress in Estoril.

Because the decisions made during the debates in Kohl's cottage have been of such importance for the further development of the EPP, I am including the complete text here. It was not intended to be read by outsiders but now that its goals have practically all been realised it is no longer confidential.

1. We need an EPP which, in keeping with its founding principles, remains open to new member parties and to cooperation, within the framework of the EPP Group, with like-minded MEPs. These founding principles are the Christian conception of the person, our European heritage (a "communitarian" Europe along with the principle of subsidiarity) and a social market economy.
2. The EPP, following its enlargement in recent years, has become a broad political movement of people's parties that come from various different geographic, historical and cultural contexts. We are no longer exclusively Christian Democratic (Christian Social) but also adhere to Conservative and Liberal values.
3. The EPP must remain a key player. If we limit ourselves effectively to the present formation, we will never achieve a majority position. We should think of new members. Among other things, a European party cannot limit itself solely to European Union parties. For this reason, we should accept within our ranks parties from candidate Member States from Central and Eastern Europe.
4. If the party Union for Europe should be founded, this would have serious consequences and be a serious threat. We have all agreed to prevent this from happening.
5. With a view to preventing the formation of the party Union for Europe, we could take a number of important initiatives in the short term:
    a. Maintain and strengthen cooperation with Conservative MEPs from the United Kingdom;
    b. Convince the RPR to honour its 1994 commitment, that is, that their members in the European Parliament will become members of the EPP Group (or possibly associate members). If they submit the request, we should make it possible for the RPR to become a member party of the EPP;
    c. Encourage the formation of a centre party in Italy with a view to renewing bonds with former Christian Democrat voters. If members of Forza Europa in the European Parliament are prepared, either as a delegation or individually, to become associated with EPP Group, the group should hold a majority vote on the matter;
    d. Invite opposition leaders from EPP parties to EPP summits as well as other parties that work together with the EPP. This will help seal political and individual bonds and rule out the formation of a competitive European party.
6. EPP summits should be held more often and with different formats. Practical preparations for the European Council should continue in their current

restricted framework, but the leaders of opposition parties within the EPP could participate in discussions and decisions on other issues.
7. We all agree unanimously to disband the EDU and to integrate it into the EPP. All the parties represented at this meeting shall leave the EDU before the end of 1998. It is vital that a clear timeline be agreed on in this respect.

Bonn, 24 March 1998

A Clear Majority

The Gordian knot was cut forever at the beginning of May. Berlusconi failed to appear at the official founding ceremony of the Union for Europe in Dublin on 7 May. On 12 May, we rejected a last-minute proposal floated by their representatives – after Forza Italia had withdrawn – namely, a sort of confederation involving the EPP, the Union for Europe and the European Liberals. This proposal was completely unacceptable to us, given that we would have to surrender our own ambition of bringing together and representing all the centre-right forces in Europe. I was granted a mandate by the Presidency to commence formal negotiations on behalf of the Group with the representatives of Forza Italia in the European Parliament.

Official membership proceedings were initiated on the following day. A "contact delegation" was set up to look after proceedings. Opponents of Forza Italia took advantage of each step to call the basis of the whole affair into question, so much so that such banal matters as the composition of the delegation had to be subject to a mandatory vote. Finally we left for Milan, the official delegation including Vice-President Hans-Gert Pöttering and Secretary-General Mario David. Berlusconi's helicopter was waiting for us when we arrived and we were flown to his villa. There and then and without further ado he personally gave us his definitive yes. The agreement was captured on photo by Mário David.

On 2 June, the Group gave its approval to formal negotiations. Forza Italia's head of delegation then declared in writing that he accepted EPP policy. Analogous to our procedure when the Conservatives joined, on 9 June a vote was held behind closed doors on the acceptance of each member individually of the Forza Europa Group. This gave the internal opposition the opportunity to mobilise itself for one last stand. And what a show of opposition it was! Resistance came as before from the Italian PPI and the Benelux members. These opponents had initially remained absent, hoping that the necessary quorum would never be reached. When it appeared that there were a sufficient number of group members present who could participate in the vote, they crowded into the boardroom to vote against the motion, again hoping that no majority would be reached. Out of a total of 136 present, 95 group members voted for and 35 against the motion, along with a few abstentions and spoiled votes. Based on this clear majority, twenty Forza Italia MEPs became members of the EPP Group.

Doubts About the Opposition

Much to everyone's surprise, the number of dissenters was quite limited. Playing to the gallery, they defended their stance as being based on principle and fundamental convictions but in fact there were material interests involved (the bigger the group, the more limited the relative influence of smaller delegations would be) – national career moves and image-building among their supporters being never far off. Only those at the heart of the Group saw through this fairly quickly. To the outside world, however, they were the "real" Christian Democrats and I – who had to bring the whole operation to a positive conclusion – ended up being almost branded as a heretic, with all the consequences involved. I bear no grudge but it did come to my attention afterwards that, ironically, the opposition members never refused invitations to participate in Forza Italia's party conferences at sunny Italian venues nor were they taken very seriously by their own supporters at home about their resistance to Forza Italia.

So Forza Italia's membership caused considerable commotion but the situation within the Group soon returned to normal. In a very short period of time it became clear that the new members were highly constructive in their approach. Criticism died out among those who worked with the Forza Italia MEPs on a regular basis, and excellent working relations were established as a result. Intrinsically there was no problem within the Group. The outside world continued to equate Forza Italia with Berlusconi, unfairly so. In fact this is still the case to this day. Little or nothing is ever heard, however, about the results achieved by, or the stability of, the second Berlusconi government, the first in Italian history to complete its five-year mandate. Nor have Berlusconi's critics given him any credit for nominating three excellent Commissioners, namely Mario Monti, the current Minister of Foreign Affairs, Franco Frattini and former MEP and two-term EPP Vice-President Antonio Tajani, today member of the European Commission.

The Rift with Prodi

Forza Italia's membership in the EPP Group was not without its consequences. Because the Group had expanded by twenty MEPs, the gap between it and the Socialists was cut to fifteen, much to the anger of their chair, Pauline Green. Within the Group, good working relations with Forza Italia played an important role in their admittance as a full member of the Party. Moreover, the collapse of the Union for Europe made the neo-Gaullists move further in our direction.

Relations within the Italian delegation of the EPP Group were totally altered. The PPI MEPs were no longer the largest Italian group in the delegation and their former opponents, the right-wing Christian Democrats, had now formed a united front with the Forza Italia representatives.

Personally, I was not in favour of pursuing national politics at the European level, but Romano Prodi left the EPP with no choice. Prodi was Berlusconi's main opponent from 1996 to 1998 when he was Prime Minister. An academic, technocrat and practicing Catholic, Prodi was very close to Christian Democracy, ideologically speaking. From a political party point of view he had PPI leanings, though was never a party member. Because he was Prime Minister of Italy, it was only logical that Prodi be present at EPP summits. That was the case on 10 July 1996 in Luxembourg. Prodi was then President of the European Council and was admired for more or less succeeding in putting Italy on track for participation in the Economic and Monetary Union. He was a welcome guest at EPP events, like the Toulouse Congress in November 1997, for example.

In early June 1998, when the EPP Group had opted resolutely to approach Forza Italia, Prodi slammed the door shut, however. He refused to come to the EPP Summit in Cardiff on 14 June and cancelled all invitations till the EPP had severed all links with Berlusconi. Despite being Prime Minister, he did not shy away from undiplomatic language and chose indeed an all-out attack. I regretted his strong reaction and tried to explain and justify the new situation from the EPP's point of view, but I could not give in to his demands. I also made it clear that the whole affair was highly delicate for me personally but that political action required resolve and perseverance: "I firmly intend to follow my personal line along the way, even if it places me in a difficult position with respect to my closest friends". The leaders at the EPP Summit in Cardiff agreed with my analysis but stressed that Prodi should continue to be invited. And yet, Prodi stayed away. His relations with the EPP would never be normal again – something that would become apparent in the years that followed.

Prodi was personally convinced that Kohl had been the decisive factor in accepting Forza Italia's MEPs. During a telephone call he told me frankly, "If that is what Helmut Kohl wants, then that is what will happen". He failed to realise that Kohl too had his doubts and as a result did not want to influence the Group's decision in any way. This was in sharp contrast with the resolute action Kohl had taken in the early 1990s to have the British Conservatives accepted as members. Even on the plane on the way to the EPP Summit in Cardiff, a heated discussion broke out with his advisers, in which he expressed his doubts about whether the EPP Group's decision was the right one. Kohl had become a close friend of Prodi and was certainly opposed to abandoning him. However, the consensus reached among the leaders at the EPP Summit in Cardiff was so strong that he fully accepted it.

The Athens Group

Working relations with Forza Italia in Parliament were good. As a result, protest from the PPI and the Benelux shifted to the national parties – proof once again that image-building among supporters was the more important consideration. On 23 June 1998 – thirty five years to the day after the foundation of the Christian Democratic Group in the Common Assembly of the European Coal and Steal Community, the predecessor of the European Parliament – the party leaders of the Belgian, Dutch, Luxembourg, Basque and Catalan Christian Democrats together with the Irish Fine Gael and the Italian PPI set up the Athens Group – after the Basic Programme of Athens (1992). The party presidents were Marc Van Peel, Philippe Maystadt, Hans Helgers, Erna Hennicot- Schoepges, Xabier Arzalluz, Josep Duran i Leida, John Bruton and Franco Marini.

The group was presided over by John Bruton, the Prime Minister of Ireland at the time and later Vice-President of the EPP. By acting together, the party leaders wished to influence policy decisions within the EPP and also to safeguard the founding principles of the Basic Programme. They were of the opinion that the Christian Democratic roots of the EPP were in danger, even though each new member party, including Forza Italia, was obliged to accept the Programme. Their actions remained limited to about four gatherings and as far I know nothing substantial ever emerged from them.

François Bayrou and the EPP Council

The expansion of our political family also had repercussions on its structure. We faced a twofold challenge: changing the Party to accommodate the growing number of new member parties (expanding) while at the same time stimulating the political debate (deepening) that was becoming increasingly important. The EPP Congresses and meetings of the party's parliament, known as the "Political Bureau" – a term that I continue to have strong reservations about – had grown considerably in size but posed no real threat to the running of these organs. This was not the case with summit meetings of EPP heads of state and government, which proved much more difficult. Following a suggestion of Kohl, EPP summits were reserved for heads of state and government, as was the case during Maastricht, for example. Kohl wanted to limit the number of participants at all costs because he found that meetings of more than twenty were not efficient. However, as the Party had expanded and because from 1998 on it was difficult not to hold summits without inviting the CDU/CSU, which did not participate in the German federal government, both government and opposition leaders were systematically invited to EPP summits. Afterwards formal distinctions were made between "statutory" EPP summits comprising Prime Ministers and Deputy Prime Ministers, party presidents and opposition leaders, and "extended" EPP summits to which representatives of associate member parties and observers could also be invited. In practice, EPP summits are invariably "extended" summits.

Party leaders who were in opposition or who were not Prime Ministers were easier to involve more closely in the running of the EPP – namely, Aznar and Juncker, who were not yet Prime Ministers at the time. Thus, initially an "EPP Council" was formed alongside the EPP Summit. The founding of the Council at the Madrid Congress in November 1995 presented me with the opportunity to give François Bayrou a more prominent place within the EPP. There was no place for him in the Presidency, and the French had to be represented in some way or other in the Party leadership. Something that I had always feared did manifest itself, however: the Council would be a stillborn child. Soon after, Juncker and Aznar became Prime Ministers and members of the Summit and they no longer needed the Council. The attempt to involve the presidents of the member parties more closely in the running of the EPP had failed. Bayrou did little or nothing to motivate the Council. On the few occasions he did participate, I always gave him the opportunity to preside over the meeting. But he never had anything prepared. On one occasion he confided to me: "I really admire you. How can sit and listen for hours to all that rubbish and try to make sense of it? I have neither the patience nor the temperament for it". And that was indeed the case. Needless to say, the Council was soon abolished.

The Leading Body Within Parliament

As Group chair, I was a member of the Conference of Presidents, the meeting of group chairs led by the President of the European Parliament. This Conference was the leading body within Parliament. The Conference dealt with political issues and if no consensus could be reached a vote was held in which each chair had a weighted vote in proportion to the number of his or her group members. Much to my annoyance, the civil servants often needed minutes on their calculators to add up the votes! Nonetheless, it was a politically varied and interesting company in which politics were discussed with a great deal of knowledge and understanding. Pauline Green represented the Socialists in the Conference. The Liberal Group was represented by the Dutchman Gijs de Vries, and later the Irishman Pat Cox. De Vries was EU Coordinator for Counter-terrorism from 2004 to 2007 and Cox was President of the Parliament from 2002 to 2004. At first, Jean-Claude Pasty sat in the Conference on behalf of the Rassemblement des Démocrates Européens, and later on behalf of the Union for a Europe Group.

The three traditional political families in Parliament – Christian Democrats, Socialists and Liberals – all have a pronounced orientation towards Europe. They have been the pioneers of greater integration. The extreme left and extreme right are each other's allies in their resistance to advancing integration. During the 1994–1999 legislative term, the centre-left (the Socialists, Greens and former Communists) held an absolute majority but I tried to reach political agreements with my Socialist colleague Pauline Green at the time. This was the case when it came to electing a new President of the Parliament. It had been agreed during my first days as Group leader that we would support the Socialist candidate for the presidency during the first half of the legislature and allow the EPP candidate to replace him or her during the second half. This produced two excellent Presidents: the German Klaus Hänsch for the Socialists, and the Spaniard José María Gil-Robles for the EPP.

To find out what kind of person I was – I was new to Parliament in 1994 – Green went to then European Commissioner Karel Van Miert, a Belgian Socialist, for advice. Being both Group chair and Party president was very unusual among the Socialists. My cooperation with Green was professional and efficient. She was a pro-European British Labour politician, although, unlike me, she conducted politics in terms of the majoritarian model. In other words, her point of reference was the British majoritarian parliamentarianism or the Westminster model. Given my Belgian background, I was more inclined to seek consensus and form coalitions. Our cooperation was vital for the proper running of Parliament, because only the combined cooperation of both our groups could provide a stable majority. This became clear from voting rounds during plenary sessions. I often had to warn our group in advance – there were those hardliners who thought they could pull off just about anything – that we could always reach a majority with the help of the Socialists. Since Maastricht, the European Parliament has been on equal footing with the Council in legislative terms for a number of issues, on condition that amendments are passed by an absolute majority of MEPs (the so-called co-decision procedure).

Attempts to do away with a number of shortcomings, like the lack of a uniform statute and salary for MEPs, or to get rid of flagrant abuses, like the infamous grant of a tax-free lump sum for travelling expenses, seemed impossible to achieve, much to my dismay. Serious attempts to regulate these issues were sabotaged time and again. Those for and against were to be found in every group. A small minority was horrified at the continuing damage being done to Parliament's reputation; the vast majority of the MEPs were little concerned by the whole affair, if at all. Had not we fought for all this in Maastricht?

Dialogue with the Orthodox Churches

One of the most important initiatives I undertook as Group leader was to start a religious dialogue with the Orthodox Churches. Even though my goal was to establish dialogue with the leadership of all major religious denominations of

Europe, the only leader that responded positively to my initiative was the Ecumenical Patriarch Bartholomew I of Constantinople, an enlightened leader with daily struggles at the Orthodox Church's seat in Istanbul, Turkey. The dialogue between the EPP and the Orthodox Churches was launched for the first time on 27–28 April 1996 in Thessaloniki and has taken place every year since. Years ago, we were considered the party of the Vatican. Today, one of the main characteristics of the EPP is its multi-denominational character. The Catholic Church retains an important influence on our basic philosophy through its encyclicals and teachings on social issues. The Orthodox Churches are known for their close links with politics and were open to political dialogue, especially with the EPP. Because the meetings with the Orthodox Churches are often held in Istanbul, increased attention began to paid to Europe's relations with Islam.

My work with the Party and the Group was determined by the Parliament's agenda. The plenary sessions lasted a whole week and were held in Strasbourg every month. That week was also full of other meetings, and this enormous concentration of activity was simply unmanageable. I would arrive on Mondays and already from early in the morning would be involved in a cycle of group meetings, gatherings of the Conference of Presidents, plenary sessions and so on. As chair of the EPP Group, I always had to be prepared, particularly when votes were being cast at the beginning or at the end of the week. As a considerable number of MEPs had either not arrived on time or had already left, a limited number of votes could tip the scales in one direction or the other.

Mondays, Tuesdays, Wednesdays and Thursdays were reserved for plenary sessions. Voting rounds were held on Thursday afternoons or on Friday mornings. Next to constituency week, during which I dedicated my time completely to the Party, in Brussels there were two weeks of committee meetings and a week of group and delegation meetings each month, along with supplementary plenary sessions. The lorries full of metal boxes that ply the roads back and forth between Brussels and Strasbourg are legendary. The new building in Strasbourg, used only four days a month, cost European taxpayers 1.5 billion Euros.

As EPP Group chair, I gave numerous speeches and met with countless dignitaries, including the heads of governments who presided over the European Council every six months. These encounters mainly took place in Brussels and Strasbourg, but not always. In March 1997, the Group attended an audience with Pope John Paul II at the Vatican for the fortieth anniversary of the Treaty of Rome. The Pope was still in good health at the time and in our conversations it was the contention, now that the Berlin Wall had fallen, that "the hour of the Christian Democrats had arrived". Other talks were of a completely different nature, such as the discussion with the American movie actor Richard Gere, who came seeking support for Tibet's independence.

On 7 November 1995, I was awarded the "medaille Robert Schuman" by our group. The medal is awarded to politicians "whose courage and vision have contributed towards the unification of Europe". In addition to my many other awards and distinctions, I was presented the Charles V Prize in 1998. The European Academy of the Yuste Foundation – named after the Monastery of Yuste in Extremadura, the monastery to which, at the peak of his fame in 1555, Charles V, Holy Roman Emperor, retired after he had abdicated and where he died and lays buried – wished to honour my many years of effort for and dedication to the process of European integration. I was presented the prize by Princess Elena of Spain during a ceremony held at the Yuste Monastery on 25 June. The prize has been awarded seven times: to Jacques Delors in 1995, to me in 1998, to Felipe Gonzalez in 2000, to Mikhail Gorbachov in 2002, to Jorge Sampaio in 2004, to Helmut Kohl in 2006 and, in 2008, to Simone Veil, the first President of the first directly elected European Parliament.

Elected Once Again

The EPP Political Bureau elected me to a new term as EPP President on 8 February 1996. I was also re-elected with ease as Group chair on 13 November 1996. Appreciation was expressed by many MEPs, but I remember in particular Luxembourg's Viviane Reding – now a two-term member of the European Commission – saying that "there was no need for an election since Martens has done so well. An affirmation by applause would suffice". Otto von Habsburg, on the other hand, who was never my biggest fan, correctly pointed out that the statutes of the Group demanded a vote but that this should be no cause for concern for me. "During all these years, we have not had a chair who has achieved so much", he acknowledged.

Yet my subsequent re-election as EPP President at the party congress in Brussels on 5–6 February 1999 took place in a very different atmosphere. In fact, it was during this meeting that I faced the consequences of supporting Forza Italia's membership. This was the first time the President and the Vice-Presidents would have to be elected by the congress, instead of by the Political Bureau. But the whole affair had been badly prepared for from an organisational point of view. Among other things, the timing of the vote was such that a large number of the delegates were unable to participate. Rinsche, the leader of the German delegation, was of the opinion that this would pose no problem but I was not at all at ease with the situation. It was as clear as daylight that the opponents of the EPP's enlargement had been mobilised. These included the Benelux parties, the Union pour la Démocratie Française (UDF) from France, the Italian PPI and the European Union of Christian Democratic Workers (EUCDW): the result was 190 votes in my favour, 61 against and 21 abstentions.

This congress also launched the campaign for the June 1999 European elections by voting in favour of the EPP election manifesto "On the Way to the Twenty-first

Century" and the 1999–2004 Action Programme. In my closing speech, I spoke of the religious pluralism in our party and did not avoid the issue of Forza Italia:

I have had to make some very delicate decisions, in particular enlarging our group to include the twenty members of Forza Italia to prevent a new political party being created; one which would be a genuine rival. I took considerable personal and political risks, which are still being held against me in my own country. We have never been a confessional party. Nor do we want to be one. But in the beginning, we were a party in which Catholic believers, in particular, played a very important role. Today, in terms of our philosophy, we have become a pluralist party with many voters, elected members and office holders who belong to the Catholic Church, the Protestant churches, the Anglican Church, the Orthodox Churches and even the Muslim religion.

The Lost Battle for Santer

The European Parliament showed its teeth for the first time during the mad cow disease crisis in 1996 and 1997. Drawing on well-founded analyses and arguments, it forced the Commission to take action. A number of EPP members played a vital role in this. The German Reimer Böge, one of our best MEPs, belled the cat during the Budget Committee meeting and rounded on Franz Fischler, Commissioner for Agriculture, at a hearing in his own committee.

The fact that things were serious in Parliament was obvious from its refusal to discharge the budget for 1996 and approve the 1998 budget. The decision was influenced by the dubious practices brought to light by the Dutchman Paul van Buitenen, a civil servant in the financial section. What had merely been a technical matter until then now became the centre of a high-stakes political battle. On top of all this, rumours were flying about other irregularities in the EU's civil service, reaching even to the highest levels of the Commission. The Socialist Group opted to take advantage of the situation and tabled a motion of non-confidence in the Santer Commission. This strategic move seemed to be designed to prevent discussions concerning the behaviour of the Socialist members of the Santer Commission. It was preferable to attack the ill-fated Santer, whom they accused of lacking resolve. But because only Socialist Commissioners were involved in the scandal, the whole affair took a strongly political turn, thus making it impossible to take action against them.

During the parliamentary sitting on 14 January 1999, a no-confidence motion received 232 votes in favour and 293 against. It followed Santer's announcement that he was setting up a special committee of independent experts to investigate allegations of fraud, nepotism and mismanagement. A report was drawn up by this group of "wise men" and was issued on 15 March, at which time the members of the Santer Commission all resigned collectively. Two paragraphs in the report sank the Commission completely. Precisely on the final page of the 135-page report, the following political judgement was stated in black and white: "It is becoming difficult to find anyone [in the European Commission] who has even the slightest sense of responsibility".

The Amsterdam Treaty provides that the Commission could be forced to resign following a two-thirds vote of no confidence in Parliament. This applies to the Commission as a whole; individual Commissioners cannot be forced to resign. Because the Commission feared that a vote of no confidence would be passed with a majority, it decided to do the honourable thing and step down. The EPP Group demanded the resignation of Commissioners Edith Cresson, the former French Premier, and of the Spaniard Manuel Marin, even though he had served well. I myself tried everything I could to save Jacques Santer and the Commission but my efforts were in vain. What an awful time that was!

With their loss in the elections and Kohl's resignation as Chancellor, the CDU/CSU found themselves in opposition. In this new situation, the German Christian Democrats were unable to reach consensus, unlike the period when Kohl was in office. This gave rise to a hotbed of protest. I am absolutely convinced that had Kohl remained Chancellor, the Santer Commission would not have collapsed. Nor was the Socialists' stance free of ambiguity. The wanted to play the role of white knight but at the same time, they tried to prevent the establishment of the OLAF, the European Anti-Fraud Office, which until then had been an internal body in the Commission. It was all very much a big mystery!

The slow death of the Santer Commission made clear the differences in style between me and Pauline Green. Whereas I sought an agreement with the Commission and pressed for talks as a result, Green insisted very strongly on so-called clear decisions. She was not prepared to reach a compromise. She managed to save face during the Commission's demise and largely succeeded in shifting the blame away from her political family, notwithstanding the fact that three Socialists – Cresson, Marin and the Fin Erkki Liikanen – were headed for the gallows.

Prodi as New President of the Commission

The resignation of the Santer Commission was on the top of the agenda at the meeting of the European Council in Berlin on 24–25 March 1999. The NATO bombings of targets in Serbia in retaliation for Serb actions in Kosovo started during the meeting. Agreement had to be reached on "Agenda 2000", the multi-annual budget for the period 2000–2006. This was a prickly matter, as was the case with all EU budgets. Because the President of the European Council, newly elected Socialist Chancellor Gerhard Schröder, needed a political victory at the beginning of the summit, he put forward Romano Prodi as Santer's successor; in so doing, he took everyone by surprise.

During the EPP Summit in preparation for the European Council, I had already asked our Prime Ministers if they were convinced that Prodi would make a good Commission President. Belgian Prime Minister Jean-Luc Dehaene answered, "Who says he will be nominated?" Wim Kok, the Prime Minister of the Netherlands, was also mistaken. He was extremely interested in the job himself but could not accept such a post since he had only recently been re-elected as Prime Minister. He was badly in need of a tactic, one that would lead the heads of government to beg him to become President of the Commission; this would appease public opinion in the Netherlands. But Schröder arranged a pre-emptive political strike with Prodi, the former Italian Prime Minister and old acquaintance of the EPP. Schröder pulled it off: Prodi was quickly chosen by the unanimous decision of the European Council. He was officially appointed to the post on 5 May, following a vote in favour by a clear majority in Parliament.

The concerns I expressed to the heads of government at the EPP Summit in Berlin later proved to be justified. In 2005, a book was published in the Netherlands entitled *De Europese onmacht* (The Powerlessness of Europe). The author, journalist Ben van der Velden, paints a damning portrait of Prodi. His work draws on confidential information gleaned from top politicians. Jean-Luc Dehaene, for example, states that "Prodi's disaster is not only the fact that he is a poor communicator with the press, he cannot even communicate with the European Council. If you do not know what precisely to say in the European Council, then you are lost. You are there on your own at dinner with the heads of government and there is not a civil servant in sight to help you. When Delors, the former President of the Commission spoke everyone was silent. His successor, Santer, kept it to the odd technical remark. Prodi could have played an exceptional role in this respect, particularly given the weakness of the European heads of government at the table. But he never managed to do so". Nor did Jean-Claude Juncker pull his punches: "Prodi is a problem. He does manage to amuse people. He has great human capacities. But he sometimes talks about things he knows nothing about. His French is limited. He does not speak any German and his English is accessible only to the initiated" (Van der Velden, 2005, p. 15). Aznar too confided in me only months after Prodi's appointment that the former Italian Prime Minister was "le chaos total".

Chapter VIII

A *Cordon Sanitaire* Around Austria

In the run-up to the European elections on 13 June 1999, I was confronted with an underhanded and devious move regarding how the list of candidates for the European Parliament was to be drawn up by my home party. The elections coincided with federal and regional elections. In the Belgian electoral system, one's position on the list is a delicate matter, as it can be vital to whether one is elected or not. Traditionally, the list comprises a mixture of experienced and newer candidates, men and women, young and old and so on. This is especially the case with the list of candidates for the European elections, as the French-speaking and the Dutch-speaking regions of Belgium each have their own electoral districts. The list of my party, the Flemish Christian Democrats, was officially approved by the party executive. But it had already been de facto decided upon beforehand by an inner cabal of the party. A candidate committee, something that any serious party would have formed, was nowhere to be found. What mattered instead was who was the object of rumour or backstabbing.

In early 1999 rumours were flying constantly in all directions. All of a sudden, a proposal was put forward within the party that there should be more women on the list and consequently, that the Minister for Employment, Miet Smet, would be an excellent choice to head the list of European candidates. Her candidacy was supported by the leaders of the Flemish Christian Democrats. For thirty years, Miet has been my favourite political companion. I had a very strong suspicion that there was something else afoot. And it was not surprising that people lacked the courage to let me know. In mid-February a decision was made in her favour.

I disagreed completely with the way I had been moved to second place. My party had suddenly broken with a long tradition of supporting its members in positions of leadership as Christian Democrats in the EPP or Parliament. This had been the case with Leo Tindemans, my predecessor as EPP President and Group chair, who had always been put forward at the head of the list during European elections. This was only obvious and he would never have accepted anything less.

It seemed that having more women on the list was being used as an excuse to replace me at the top of the party's list. I was paying an extremely high price for allowing Forza Italia into the EPP. I was also the victim of a smear campaign that had been running for years; namely, I was being blamed personally for the fact that the Belgian national debt had increased considerably during the 1980s. However, it was my government who took the first drastic reform measures in 1982, following the successive shocks of the oil crisis. On top of this, there were repeated insinuations that a once-divorced and now remarried father of young children could never head the list of candidates of a Christian Democratic party.

Though there were those who tried to convince me that all of this was untrue, I stood firm and refused to accept second place. Many failed to understand that I had chosen to resign early from Parliament on grounds of principle. And yes, there was still a lot of work be done in developing and consolidating the EPP Group. And yes, I had been proposed for the presidency of the Parliament – a position I would have accepted with utmost conviction and enthusiasm. Yet my departure from Parliament did not weigh too heavily on me. I had reached the conclusion that combining the task of party president with that of group leader was inhuman and impossible – if one wanted either done properly, that is. So my leaving the Group was liberation for me, and was also good for the Party because I could focus my efforts fully on the future of the EPP. In fact, a party of that size and potential needed a full-time president.

Friendship and Solitude

The reason I was not given the top spot on the list for the 1999 European elections was political. They wanted to prevent me from heading the list, in anticipation of obviously new trends in Belgian and European politics. The reasoning was that without new people and innovation my party would irrevocably crumble. On the European front, the EPP stood an excellent chance in the elections. So I would become the symbol of victory in Europe and my opponents would be the losers in their own country. This could never be allowed to happen, because that would have meant my return to national politics.

All this in fact came about on 13 June 1999. My party suffered a historic defeat and stepped down after decades as head of government, moving in humiliation to the opposition. The EPP did indeed pull off a spectacular victory in the European Parliament and after many years became the strongest group once again. Five years later, it emerged that the need for a woman to head the list had only been a pretext. When the list of Flemish Christian Democrat candidates was drawn up for the 2004 European elections, Jean-Luc Dehaene was put forward without discussion to head the list. I hold no grudge against Miet Smet, who headed our party's list instead of me in 1999, since nine years later she became my wife and loving companion.

The great difference between democratic and non-democratic countries is that the latter have no respect for the privacy of their citizens. They want to know the convictions, ideas and feelings of their subjects and are not afraid to use illegal methods to find out. This is also a characteristic difference between democratic parties and non-democratic parties. For this reason, I have always refused to make public anyone's private life, not even that of my strongest opponent. Had I lived with a judicial system that works well at protecting the privacy of its citizens, like France's, for example, I would have taken people to court years ago. When a journalist asked me if I would do the same thing if I could live my life over again, my answer went like this: "I would hesitate, very much so. I have sacrificed much and expect no thanks in politics. But there have been deep wounds".

In my farewell letter to European voters I wrote the following: "I have never belonged to the establishment of the Flemish Christian Democrats but I hope, given my experience, that I will be able to contribute something to the party. In the future, I will dedicate myself hundred percent to my EPP presidency. Through considerable personal effort, I have achieved results in this position in the past. Friendship, belonging, idealism and so on can be among the most beautiful experiences in a person's life. Unfortunately, they are accompanied by loneliness and can never compensate for it".

As a politician there is nothing worse than being left out of decisions that concern one personally. It is not surprising, therefore, that from time to time a marginalised politician responds bitterly to questions put to him or her by journalists. Information is power. Excluding someone from the normal channels of information is often tantamount to professional murder. During a state visit to Belgium in 1983, President Mitterrand hit this very nail on the head: at a dinner at the French embassy, he told King Baudouin that communist parties always isolate their leaders politically; no one ever pays them a visit; they are denied any form of information. He gave the example of Hungary after the 1956 revolution. *Kaltstellen*, literally, "making cold" or "freezing", is what it is called in German. People who are always talking about the principles of democracy should realise this.

My Heritage

As soon as it became known that I would not run in the 1999 European elections, it was obvious that vice-chair Hans-Gert Pöttering, who had been an MEP since 1979, would succeed me as EPP Group chair. I had developed very good relations with him over the years. Knowing that he was someone who knew the Group inside out gave me confidence in the future.

Pöttering's term as chair turned out to be much better than some had believed – or feared – it would be. Initially there had been considerable scepticism about him: he is a German – still a sensitive issue for some – and in broader circles little was known of him at the time. He is considered a diplomat and a man of consensus; I share with him a deep commitment to Europe. Though I became totally involved in my European mandate, Pöttering made sure he never lost touch with national politics and the media. I could have learned a thing or two from his attention to public relations. "Schade dass es keine Bilder gibt"; "It is a pity there are no pictures", he complained after he and I had just met President George W. Bush at the White House in July 2005. However, he was less careful in his political statements. He could afford to be as he was spokesperson of the largest delegation in our group.

The EPP emerged from the polls as the biggest winner of the 1999 European elections. We were the largest group for the first time since the start of direct elections to Parliament in 1979. This was the combined result of a few striking electoral victories – that of the CDU/CSU in Germany and the Tories in the UK, among others – along with an enlargement of the Group during my presidency. Following the battle for the polls in 1999, our group was 232 strong, compared to 157 for five years in the previous term.

Given the fact that the German Christian Democrats, led by their new President Wolfgang Schäuble, had carried more than half of the German seats in Parliament (53 of the 99) barely one year after their defeat in the German Federal elections, the victory was all the more resounding. The good result achieved by Edmund Stoiber's CSU was clearly a part of it. The Tories pulled off a tour de force and doubled the number of their seats, rising from 18 to 36 following the introduction of the proportional system.

William Hague became Tory leader following John Major's defeat in the UK elections in 1997. Their attitude then changed completely, because the party's Eurosceptics were given a free rein. Their rapprochement with the EPP came to a halt. Even worse, those in favour of separation from those damned continental federalist Christian Democrats began systematically to gain ground. The British Conservatives also tried to capitalise on their strengthened position within the EPP Group.

During the EPP study days in Malaga, I met Silvio Berlusconi, leader of the Italian opposition and newly elected MEP, Prime Minister José María Aznar, CDU President Wolfgang Schäuble and Tory leader William Hague for an informal dinner. They seemingly had already agreed amongst themselves to change the name of the Group to the "EPP-Conservatives". I was flabbergasted when I discovered this during the course of the dinner. "Are you out of your minds?" I said straight away. "If we do this it will mean the end of the EPP in many countries!" This explicit reference to the Conservatives was unacceptable to many members of our party, and I knew it would lead to a rift. A party name like that would unnecessarily irritate vast numbers of our activists. The expansion of the EPP had been compromised and was in danger of collapsing.

Fortunately, disaster could be averted. I remembered that the Tories had been members of the European Democrat Group prior to joining the EPP Group. As a result, I suggested that the name of our group be extended from EPP to EPP-ED, ED referring to European Democrats. Though I was, in fact, opposed to a name change, this seemed like the only way to save the day. The new name, EPP-ED Group, was ultimately accepted by all.

From Bad to Worse

However, the Malaga agreement never led to a stable understanding between the EPP Group and the British Conservatives. Matters only went from bad to worse. In the course of the following ten years, relations kept sliding steeply downhill, for after Hague came Ian Duncan Smith, who put our agreement in doubt once again. He made that clear to me during our meeting in London on 22 April 2002. Under Duncan Smith, the Eurosceptic faction of the party, which had voted against the ratification of the Maastricht Treaty, was now in the driver's seat of the party. Smith continued to insist on a far-reaching special status for the Tories within the EPP-ED Group.

They already had been given this to a certain extent, because since the extension of the name, they had an extra vice-chair, could elect their own whip and so on. Particular efforts were made within the Group to ensure that the British Conservatives were clearly represented. The vast majority of the British MEPs also wished to remain within the EPP-ED Group because the costs would far outweigh the benefits if they were to leave us.

Nor was Duncan Smith's successor, Michael Howard, too impressed by our cooperation. But he was intelligent enough not to be absent from EPP summits. Tensions subsided when he took over the leadership of his party. The EPP Summit in Paris on 4 December 2003 normalised relations again, but Howard, nonetheless, proved a cool and calculated partner. That is why he managed to get his way in the Group. Following numerous rounds of negotiations, the Group regulations were amended, thereby allowing the Tories to develop and promote their own points of view on EU affairs within the Group. This was the price that seemingly had to be paid to keep them on board ahead of the European elections and hence increase our chances of remaining the largest group in Parliament.

Allowing such autonomy within the Group was a highly risky step to take. According to those in favour, it was merely a confirmation of what in fact was already the case. But opponents felt it was one bridge too far, because up until then the Group had never at any time during its expansion conceded any ground on pro-European positions. Moreover, there was real fear that parties from the new Member States might choose a European Democrat section of the EPP-ED Group rather than the EPP section. The German delegation closed ranks in favour of the change. This reform was led by Group leader Hans-Gert Pöttering and Secretary-General Klaus Welle, who had moved to the Group in 1999. The Group announced the agreement on 31 March 2004 but only a year later we paid dearly for it.

Déjà-vu

The new Tory leader, David Cameron, put his party's membership in the EPP-ED Group into question yet again in late 2005. In his campaign for the leadership of the Conservative Party, he had promised his supporters that he would pull out of our group and form his own group as soon as possible. In a letter dated 6 December 2005, I set out the history of our turbulent relations, reminding him of our various prior agreements and also of the repeated efforts that had been made to accommodate the Conservatives' wishes. I had been hoping for better cooperation and invited him to the following EPP Summit. During the summit, the EPP heads of government and party leaders decided that if Cameron were to go ahead with his threat to leave our group, we would refuse any form of bilateral contacts with him and his party. This was not unimportant, given his ambition to become Prime Minister of the UK. Who then would be his European partners on the day he became Prime Minister? But none of this was keeping him awake at night for the moment. His threat to pull all Tory MEPs out of the EPP-ED Group has failed to materialise till now. But the prospect of setting up a new Eurosceptic group is still in the cards. This was clearly visible in Cameron's interview with the *Daily Telegraph* on 6 March 2007. At the beginning of that year, he set up the so-called Movement for European Reform together with the Czech Občanská Demokratická Strana (ODS, Civic Democratic Party) MEPs who also sit in the ED section of our group. This effort is part of the preparation to form a new group after the 2009 European elections, but it continues to be limited to these two parties and is not yet enough to achieve the necessary critical mass.

I have done all within my power to improve relations with the British Conservatives. In this respect, I defended the distinction between the Party and the Group in changing the statutes of the EPP Group. I went to the trouble of explaining this in detail because many – some in good faith, but many more in bad faith – continued to criticise the EPP for having sold its soul for a mess of pottage. This is clearly untrue. The EPP has never compromised its principles or changed its rules for the sake of the Tories or any other member party for that matter. Let us not forget that the Party is the ideological bedrock, the cornerstone of our European political movement, motivated by the commitment of our national member parties and its leaders. Our parliamentary group, on the other hand, similarly to national parliamentary groups or factions, is a more fluid and flexible political constellation. Whereas the Group is free to ally itself with the Eurosceptic Tories or the ODS, the Party continues to steer a consistently pro-European course.

The difficulties we continue to experience with the Tories stem from their fear of, aversion to, or even abhorrence of, federalism. Being part of a transnational European party that has federalism as one of its central principles is too much for them. It is clearly problematic that the British have little experience of federalism within their own borders, although the recent devolution in Scotland and Wales has ushered in a change in this respect. More to the point, their stance seems to be a consciously created and maintained misunderstanding. The Eurosceptic diehards within the Conservative Party perceive the United States to be their big brother. Some even want to become the fifty-first state and leave the EU. The irony is that many pro-Europeans regard the US as a highly sophisticated federation.

I am not in favour of the notion of a United States of Europe. Within the context of European integration, it has never been the intention to absorb the various nations into one European superstate. The EPP does not want this either, and yet this is something that Eurosceptics have falsely accused us of, often deliberately. Exaggerated Europhilia is an image strongly propagated in the tabloids, so much so that it has become almost generally accepted as a given in British public opinion. It will require much courage and leadership to make clear that federalism is not to be equated with centralism; on the contrary, federalism provides a safeguard for devolution and decentralisation, on the basis of the principle of subsidiarity.

Finally the French

In its expansion strategy the EPP continued its search for stronger representation in France. Despite the demise of Robert Schuman's once-successful Mouvement Républicain Populaire (MRP), the tradition of Christian Democracy continued to live on in France. During the formative period of the EPP, I met Jean Lecanuet, President of the Centre des Démocrates Sociaux (CDS). I was also in regular contact with Alain Poher and Pierre Pflimlin, both of whom have been President of the European Parliament. The CDS was one of the founding members of the EPP. Nothing changed in 1978 with the foundation of the Union pour la Démocratie Française (UDF), which included the CDS. Following Pierre Méhaignerie, François Bayrou became President of the UDF in 1994. The problem was that the CDS remained a relatively small party and only accounted for one part of the centre-right in France. An attempt to include the Liberal and Republican wings failed in late 1991. Only individual MEPs, including the former President of France,

Valéry Giscard d'Estaing, and Alain Lamassoure, made the move from the Liberal to the EPP Group. The neo-Gaullists of the Rassemblement pour la République (RPR) were also being considered as potential EPP members.

The UDF functioned as a bridge builder at the time. The RPR had formed a common front with the UDF for the 1994 European elections, led by the Christian Democrat Dominique Baudis. Both parties had promised that their MEPs would join the EPP Group. An agreement had been reached on this by the then party presidents, Jacques Chirac of the RPR and Giscard d'Estaing of the Parti Républicain (who, like the Christian Democrats, joined the UDF). The EPP Summit of 8 December 1993 accepted this proposal. But the RPR did not keep its word – for which Nicolas Sarkozy later apologised to me. Their MEPs maintained their own group, the Rassemblement des Démocrates Européens. By doing so, they had their own chair, Secretary-General, staff and so on. Following Forza Italia's move to the EPP and the failed attempt to form a new European party, Sarkozy announced on 30 May 1999 that all elected RPR members would join the EPP Group. "You won", were the words addressed to me in Strasbourg by RPR President Philippe Séguin, who had worked hard to set up the Union for Europe. Under Alain Madelin's leadership, MEPs of the Démocratie Libérale, which was also part of the UDF, joined the EPP Group.

In keeping with the commitments made during the "cottage" talks in Bonn on 24 March 1998, the rapprochement between the EPP and the neo-Gaullists came about because of various meetings between Chirac and me, Kohl and other EPP leaders. Important personalities within the RPR had given the party a more pro-European profile. As Minister of Foreign Affairs, Alain Juppé played a considerable role in all of this. It was obvious to us that our efforts to broaden the party had to include the RPR. After many ups and downs, things finally became clear in January 2001, with the presence of the then RPR President Michèle Alliot-Marie at our congress in Berlin. Later that year, the RPR was accepted as a full member party of the EPP.

During one of my meetings with President Chirac, I was accompanied by my colleague and Christian Democrat MEP, Nicole Fontaine. On leaving the Elysée, Chirac asked me, "Is Nicole in the running for President of the European Parliament?" This was a strange question, as we had not even begun to think of possible candidates for the presidency of the Parliament at the time. Nevertheless, Nicole Fontaine was elected President of the Parliament in 1999 and held the position until 2001. I often ask myself what would have happened had things run their course and had I remained an MEP.

The Union pour un Mouvement Populaire (UMP) was founded in the aftermath of the 2002 presidential and general elections in France and comprised the RPR, the Liberals, the Republicans and a number of Christian Democrats, including Pierre Méhaignerie, Nicole Fontaine and current European Commissioner, Jacques Barrot. Under François Bayrou's leadership, the UDF, which included Christian

Democrats, did not join the movement. Since then, the EPP has gained a steady partner in France and our lines of communication reach to the very top of French politics. Former Prime Minister Jean-Pierre Raffarin played a decisive role in this. I had already known him as a member of our group. He was the first French Prime Minister to ever participate in our activities. The current French President, Nicolas Sarkozy, is also strongly committed to the EPP: at a crucial moment in 1999, he pushed his MEPs to become members of our group. His participation at our congress and summit meetings has given them a particular brilliance.

I have always cherished the French language and culture. I have watched French television for years, French is my second language, Paris is close by and, because of the many family holidays I have spent in the France countryside, it has become quite familiar to me. Of course, I also follow the ins and outs of French politics closely. I have noticed time and again that in contrast to parliamentary systems, in a presidential system it is a difficult task to form and maintain coherent parties, for everything depends on the fate of the candidate for president, *le présidentiable*.

Exit the EDU

One important aspect of our rapprochement with the neo-Gaullists was the relationship between the EPP and the European Democratic Union (EDU). The point of departure was that a certain "rationalisation" should be reached regarding European transnational organisations addressing centre-right parties. Some EPP member parties wanted nothing to do with the "right-wing" EDU, whereas some EDU member parties, like the RPR, for example, had hardly any affinity at all with the "Christian Democrat" EPP. Thankfully, the majority of centre-right parties belonged to both organisations and ultimately understood the need to have only one strong European organisation.

Given the growing importance of transnational organisations and their presence in the Member States, the situation of having two competing organisations became politically unsustainable. A simple merger seemed out of the question because not all parties were members of both organisations. If the EDU were simply to be absorbed by the EPP, parties from non-EU countries like Turkey, for example, would find themselves left out in the cold. But for us it was clear that the EPP would form the hub of integration for the two organisations. This formed the spirit of the decisions taken during the "cottage" talks in Bonn. The Austrian ÖVP had played a prominent role in the EDU since its foundation. The EDU Secretariat was located in Vienna. Aloïs Mock was their chair until 1997 and was succeeded by fellow party member Andreas Khol. Both chairs had steered clear of the Eurosceptic tendencies that existed within the EDU and invested most of their time in building relations with parties in Central and Eastern Europe. Common initiatives were undertaken by the EPP and EDU, but they were never really heartfelt. There was a great fear within the EDU of becoming totally absorbed by the EPP.

The Fin Sauli Niinistö and the Austrian Alexis Wintoniak became EDU chair and Secretary-General, respectively, in 1998. On the occasion of the EDU's twentieth anniversary, the document "Towards a Majority" was drawn up, in which a plea was made for a single organisation. As a result of this plea, the presidencies of the EPP and the EDU met at Cadenabbia on 20 September 1998. In the EDU document "Restructuring the Co-operation of the Christian Democrat, Conservative and Like-Minded Parties in Europe", it was proposed that the committees, working groups and secretariats of the two organisations be combined. This resulted, among other things, in the transfer of the EDU Secretariat from Vienna to Brussels.

When Wintoniak resigned in 2002, the curtain fell on the EDU de facto. In a symbolic move, the final act took place at the EPP Congress in Estoril in October 2002, when the merger of the EDU with the EPP was formalised. Incorporating the EDU tradition, the EPP modified its statutes to accept, as associate members and/or observer members, parties from countries who were at least members of the Council of Europe, and not just those from EU Member States, which was the original focus of the EPP. This allowed a host of new parties to be involved in all of the EPP structures, including the EPP Summit. This is why today the EPP Summit includes, in addition to our EU Prime Ministers and leaders, leaders from Norway to Albania and from Belarus to Turkey and even from Georgia.

A *Cordon Sanitaire* Around Austria

Among our new member parties the Austrian Christian Democrats were the most familiar with the EPP. During the 1990s, the ÖVP became a permanent fixture, mainly as a result of the outspoken commitment to Europe of its leader, the former Austrian Foreign Minister Alois Mock, and his successor, the two-term Chancellor Wolfgang Schüssel. This was also evident in the work carried out since 2002 by the former party leader and Vice-Chancellor Erhard Busek as coordinator of the Stability Pact for South-East Europe. The active pro-European commitment of these Christian Democrat leaders made Austria a country comparable, in its European orientation, to those of the Benelux.

All this seemed to come to an end in 2000 when the ÖVP formed a coalition government with Jörg Haider's Freiheitliche Partei Österreichs (FPÖ). Following Schüssel's appointment as Chancellor on 4 February, the heads of state and government of the other fourteen Member States announced sanctions against Austria. The European Treaties did not provide for such sanctions; strictly speaking, the sanctions were unlawful. The Commission and the Parliament did not endorse the sanctions. The position adopted by the various governments was not devoid of self-interest, either. They were trying put on a brave face by laying the blame at the feet of Austria. The high degree of animosity between those in favour of the *cordon sanitaire* and those against it led to a deep crisis in the EU.

This was not without its consequences for the EPP. MEPs and the Commission tried to express some understanding for the difficult situation the ÖVP found itself in. However, some of our own member parties were looking for a confrontation. This opposition to the situation in Austria was driven mainly by local interests. There are two important points that should be remembered in this respect. The EPP has never worked with extreme-left or extreme-right parties or groups at the European level. Moreover, we are unaccustomed to making declarations regarding how member parties should conduct their internal politics. In the light of these principles, it was very difficult if not impossible for the EPP to pass judgement on the situation in Austria.

The Spanish Prime Minister, José María Aznar, invited leading politicians from the EPP and the EPP Group for informal talks in Madrid on the weekend of 5–6 February 2000. He himself was completely immersed in an electoral campaign designed to take over the centre of the political spectrum in Spain. The coalition with the extreme right in Austria would seriously compromise him politically. His reaction has to be understood in this light. At a dinner at the Prime Minister's residence, La Moncloa, Aznar suggested that the ÖVP be suspended from the EPP. We were able to prevent such a radical decision from being taken. A way had to be found to treat the ÖVP honourably. The situation was delicate and tense and so full of risk that I decided to cancel the EPP Summit in Lisbon scheduled for 23 March of that year.

Keeping a Low Profile

The statutes of the EPP at the time stipulated that at least three member parties were needed to move a vote of expulsion against another member party. The French-speaking Belgian Christian Democrats, the Italian PPI and the French centrists of the UDF took the initiative. For them too the decisive argument proved to be the battle against the extreme right on the home front. The request was first discussed by the Political Bureau on 10 February 2000. In compliance with the statutes a period of reflection of at least one month was needed before a decision could be taken.

A compromise was tabled at the following meeting of the Political Bureau on 6 April. The EPP would set up a monitoring committee that would draw up a report on the political situation in Austria. During that period the ÖVP would refrain from participation in EPP policymaking. While awaiting a definitive decision, the ÖVP's membership would be temporarily suspended. The committee comprised Wim van Velzen, Vice-President of the EPP, and two delegates from the EPP Group, Gerardo Galeote from the Spanish Partido Popular and Hartmut Nassauer from the German CDU/CSU. This compromise was accepted by all and the demand for ÖVP's expulsion was withdrawn.

The monitoring committee presented its report to the Political Bureau on 6 June. Because the political programme of the Austrian government did not contain any incriminating elements, the three EPP "wise men" recommended ÖVP's immediate rehabilitation. The committee found that policymaking has to be evaluated according to its results and not in terms of declarations made by a party leader, in this case Jörg Haider, who was not even a member of the government. In addition, they denounced the hypocrisy of the Socialists and formulated a number of recommendations with regard to internal operations at the EPP. The Political Bureau backed the report's conclusions. As President, I proposed that the report's conclusions be accepted, which was agreed following a few amendments to the document. Acceptance fell short of unanimity by only two votes. Following a lapse of two months the ÖVP could once again participate in EPP policymaking. I welcomed Wolfgang Schüssel as the new Austrian head of government at the EPP Summit at the end of June. The crisis seemed to have blown over.

However, the whole affair could not be tolerated by the French-speaking Belgian Christian Democrats, the Italian PPI and the French UDF. Led by UDF President François Bayrou, a few disgruntled MEPs set up the "Schuman Group" on 23 June 2000, the name referring to one of the founders of the EU, Robert Schuman. The leaders of the Schuman Group, who stemmed mainly from the French, Benelux, Spanish and Italian delegations, formed their pressure group under the pretext of upholding the basic principles of Christian Democracy, which had been put under considerable pressure by the positions taken by the EPP and the EPP Group, they said. Their meetings were limited, however, to agreements about votes on day-to-day problems in the Group and in Parliament. Their efforts were poorly coordinated and their numbers remained limited to only 40 of the 230 EPP Group members, at the very most. Not one of their initiatives in Parliament came to fruition. The Schuman Group began to disintegrate during 2003 and 2004 when the PPI and the UDF turned their back on the EPP and began talks with the European Liberals. In April they announced the establishment of a transnational party called European Democratic Party, and after the June European elections they allied themselves with the European Liberals in the European Parliament as part of the Alliance of Liberals and Democrats for Europe, or ALDE Group.

An Example and a Pioneer

Once the EPP had put its house in order, I made efforts to have the Austrian government rehabilitated within the EU. In the run-up to the European Council meeting in Feira, Portugal, in June 2000, pressure was brought to bear on heads of state and government who had continued to cling stubbornly to the *cordon*. The EPP achieved a remarkable victory at the European Council, for it adopted the reporting formula used by our monitoring committee. One of the three "wise men" of this new EU committee was former Commissioner Marcelino Oreja. In early September they presented their report, which exonerated the Austrian government from all accusations and speculation. The *cordon* around Austria was lifted unconditionally on 12 September. One can rightly ask in retrospect what the real purpose was of all the fuss.

During this turbulent period, the CDU/CSU was unable to play the role of stabiliser in the EPP because, as the opposition in Germany, they had to deal with their internal crisis resulting from the *Spendenaffäre*, or financing scandal. At that time, Aznar was the only EPP head of government from a large Member State. But the predominance of Socialist heads of government in the European Council proved a determining factor in the way Austria and the EPP were addressed. Ultimately, the ÖVP was far from weakened when it emerged from this delicate period. On the contrary, the low profile adopted by the ÖVP and their leader Schüssel had, in fact, helped considerably in diffusing the crisis.

I do not wish to assert that the approach would be suitable for everyone, but Wolfgang Schüssel was proved right, politically speaking, in his strategy of challenging the extreme right via a coalition. He did not make compromises and undertook no initiative that was not within international and European rules of law. Schüssel, one of the most active participants at EPP summits, was tipped as a candidate for President of the Commission in 2004 but was faced with the continuing prejudice of some heads of government and state.

Disaster at Nice

The European Council needed a quick way out of the Austrian crisis, because of the potential negative effects the crisis could have on the Intergovernmental Conference on the new Nice Treaty, which was taking place at the time. All Member States, including Austria, had a right of veto. Those present certainly did not want things to go that far, as the new treaty was designed to prepare Europe for enlargement and the subsequent inclusion of new Member States from Central and Eastern Europe – something that had not been achieved in Amsterdam. An urgent agreement now had to be reached on those so-called leftover issues that remained on the negotiating table. In other words, Nice was a last-chance treaty.

However, the agreement that the heads of state and government reached after a marathon session was decidedly second rate. The solution reached with regard to the outstanding issues – the number of members in the Commission, the proportion of votes in the Council, the definition of a qualified majority, the maximum number of MEPs – was full of anomalies and largely missed its target. Nice provided no real solution to the challenge of enlargement and was a sad failure as far as any deepening of the EU was concerned. An EU of twenty-five Member States was not ready to be run efficiently, democratically and as closely to its citizens as was desired. Nonetheless, Spain and Poland were delighted by the number of votes they had in the Council. They would defend Nice to the bitter end – *Nice ou la mort!* was a slogan regularly heard in the Polish Parliament. It became clear yet again that an Intergovernmental Conference, limited as it was to representatives of national governments, was no longer efficient. The debate about institutions, when it did take place, was totally dominated by the interests of the Member States. Nobody had shown any concern for, let alone defended, the general interest of Europe.

Fireworks at the EPP Congress

Immediately after the Nice Summit, the EPP held a congress in Berlin on 11–13 January 2001. Much to the displeasure of the EPP heads of government, Nice came under heavy fire at the congress. My address contained unveiled criticism of the new treaty. But with a view to repairing the shortcomings of the Nice Treaty, following a suggestion made by our group, I proposed that a convention be held comprising not only government representatives but also members of national parliaments, the European Parliament and the Commission. There had been a precedent: the 2000 Convention led by the former German President Roman Herzog, which, to everyone's satisfaction, had reached a consensus on the Charter of Fundamental Rights of the European Union. Work had to be started again but could not be left in the hands of diplomats from the Member States. Once the convention was agreed to at the European Summit in Laeken, the EPP Group ratified the Nice Treaty. Despite the sharp criticism, we felt that we could afford a new crisis. The Nice Treaty was tolerated only because we knew it would be reviewed.

At our Berlin Congress we defined the EU as a "Union of Values". The Basic Programme of Athens states that the EPP is a "party of values". It soon became apparent that this union of values was very important. Once we launched the concept and the ideas behind it, it soon found its way into the official documents and discourse of the EU. It is striking how one can work on a document for months only to discover suddenly that the title or concept turns out to be more important and engaging.

Since the enlargement of the EPP in the 1990s, the Party was no longer predominately Christian Democrat in composition but nonetheless remained so in its goals and message. The EPP remained unambiguously faithful to a federal EU.

The reference to a "United States of Europe" – a concept that hails from the very beginnings of the EPP and which was rendered obsolete by developments within the institutions – was dropped, however.

Verständnis mit Schäuble

Immediately following the Berlin Congress, Wolfgang Schäuble and I drew up a new proposal for the institutions. The working group that we chaired together was a prime example of our good working relationship. Following a request from the new leader of the CDU, Angela Merkel, I had asked Schäuble if we could draw up together the congress document for Estoril. Following the September 1998 electoral defeat of the CDU/CSU, Schäuble had performed wonderfully as CDU party president and group chair in the Bundestag and, among other things, by winning an impressive number of regional and European elections. If he had remained CDU President, he would have become Chancellor in 2002. But once he was drawn into the political tsunami caused by the financing scandal, there was little room left for him at the top. Personally, I got on very well with Schäuble. He is one of the people I always look up when I am in Berlin. We share a *communis opinio* both in terms of the future of Europe and all it involves, and in some respects our congress document, "A Constitution for a Strong Europe", reflects this.

When Belgium held the Presidency of the EU, Prime Minister Guy Verhofstadt had a declaration passed on the future of Europe during the Laeken Summit in December 2001. This declaration formed an answer to Nice and was also a great step forward. It provided for the setting up of a convention led by Valéry Giscard d'Estaing and two Vice-Presidents: former Belgian Prime Minister Jean-Luc Dehaene and former Italian Prime Minister and professor of constitutional law Giuliano Amato. Work on the Convention on the Future of Europe, as the convention was known in full, began officially on 28 February 2002.

Brok's Energetic Group

Thanks to our working group, the EPP was well prepared for the Convention. Through the document that was approved during our congress at Estoril on 17–18 October 2002, we were able to give form to our proposal and to take a fundamental step towards deepening the EU after Nice. At the same time the Estoril document formed the basis for our commitment to the Convention.

In order to maximise our political impact, we set up an EPP Group inside the Convention itself. I asked one of our heavyweights, Elmar Brok from Germany, to lead it. He did so with verve, personifying and enlivening the presence of the EPP within the Convention. Our Convention group was perhaps one of the most energetic and vital we have ever known. Our meetings were of the highest standard, as can be seen from the excellent report drawn up by the French MEP Alain Lamassoure regarding the distribution of competencies between the Member States and the EU. One of the most striking persons present was Helmut Kohl. He had considerable moral authority and in each of his interventions continued to stress the importance of the role played by each member of the Convention. Proceedings were even quite relaxed at times. During our meeting at Petersberg near Bonn, he interrupted John Bruton in the middle of a very complex exposé. "Das ist absurd, absurd, absurd!" he remarked. "Thank you very much for your positive comment!" replied Bruton laconically.

No fewer than seven of the thirteen members of the Convention Presidium belonged to our political family, which certainly increased the EPP's influence: next to the President, Valéry Giscard d'Estaing, and one of the Vice-Presidents, Jean-Luc Dehaene, there were the French European Commissioner, Michel Barnier, the Spanish Minister for Foreign Affairs, Ana de Palacio, the former Irish Prime Minister, John Bruton, the former Prime Minister of Slovenia, Alojz Peterle – the only representative in the Presidium from the new Member States – and the Spanish MEP Iñigo Méndez de Vigo. Méndez de Vigo is probably the least known of the seven. He is, nonetheless, a highly competent expert on the institutions and an extremely committed Member of the European Parliament.

Among the most important preparatory rounds were the informal meetings of EPP leaders on 9 September 2002 at Porto Rotondo, the small port in Sardinia where Berlusconi has a holiday home; on 19 June 2003 at the Kassandra peninsula, near Thessaloniki, where we reached an agreement on a draft of the Constitution; and on 4 December 2003, a meeting organised by Prime Minister Jean-Pierre Raffarin at his official residence, Hôtel de Matignon.

Constitutional Ambitions

That a consensus was ultimately reached within the Convention on a Draft of the European Constitution and that a last-minute agreement was reached on a Treaty for the Establishment of a European Constitution – leaving aside whether the term "constitution" was an opportune one – was all very much in line with what the EPP and I had been in favour of for years. In 1990, even before the Maastricht Treaty, we had been discussing the idea of a constitution within the EPP and had continued to do so ever since. Neither have I ever hidden my belief that a day will come when a European Constitution replaces all the existing treaties, but the pursuit of a constitution should never be an end in itself.

That so much ground was covered, at least as far as the drawing up of the draft of the European Constitution was concerned, is largely thanks to the presidency of the Convention. Choosing Giscard proved crucial for the success of the whole enterprise. Jacques Delors was also a possible candidate and would have done just as well. But Chirac opted for Giscard for internal political reasons. As a convinced European, he had the intelligence, standing and qualifications to make a success of it. There are few with that same degree of stature who could emulate him. He cherished great ambitions and often liked to compare the Convention to the Philadelphia Convention during which the Constitution of the United Sates of America was drawn up at the end of the eighteenth century. This was probably a far-fetched comparison. Paradoxically, it was France who voted against the Constitutional Treaty in a referendum held in early May 2005. Giscard was very disappointed. "Le système européen fonctionne mal. La France n'a plus de projet européen. Les Européens n'ont plus de projet européen" (The European system functions badly. France no longer has a European project. Europeans no longer have a European project), he declared several months later.

The credit due to the Vice-Presidents of the Convention is at least as great, even though their positions and actions were sometimes driven by pragmatism or, shall we say, realism. It is well known that Dehaene and Amato worked feverishly behind the scenes to reach a compromise, which came about laboriously, given the great differences of opinion that somehow had to be overcome. Much to my disappointment, one of the elements abandoned in the compromise was the *invocatio Dei*, the reference to God – or, in a watered-down version, the role played by Christianity in the becoming of Europe. I continued to make a case for the wording by referring to the Polish Constitution: "The values of the EU comprise the values of those who believe in God as the source of truth, justice, goodness and beauty, and of those who do not share such a belief, but who through other sources accept and defend these values".

Success always has a parent-like quality to it; by contrast, failure is an orphan. And yet I believe that the role played by the EPP was crucial to the success of the Convention. The following can be added to the list of EPP achievements: the scrapping of the three pillars in favour of the distribution of competencies between the EU and the Member States; the creation of a European Minister for Foreign Affairs, who is also a Vice-President of the European Commission, something in which Dehaene played an important role; an extension of and refinement in the application of the principle of subsidiarity, a decisive contribution to this being made by Méndez de Vigo; the double majority in the Council; more transparency in policymaking; new competencies for the European Parliament, albeit fewer than expected; and the appointment of a President of the European Council, with the prospect that in time, this position would be merged with the President of the European Commission. However, we should never stop envisaging, for the future, the election of the President of the Commission by the European

Parliament, or possibly even by direct vote of the electorate. It is a hopeful sign that already in 2004 – even before the existence of a definitive draft of the treaty – the results of the European elections were taken into account in the appointment of the President of the European Commission.

Chapter IX

"He Did Not Take the Fruit of Those Who Had Planted the Tree ..."

Since the five-year terms of the European Parliament and Commission coincided, there was lively speculation about who would lead the Commission after the 2004 European elections. Romano Prodi no longer stood a chance of a second term as Commission President – even he did not think so. In the media and in countless meetings speculation reached fever pitch. We thought back to 1999. After those elections, the EPP Group had once again ended up as the largest group. Yet our victory had had no impact whatsoever on the composition of the Prodi Commission, let alone on who would be the Commission President. But in 2004, the proposed European Constitution determined that government leaders in the European Council would have to take into account the results of the elections.

In February, a European politician telephoned me and asked me to contact Belgian Prime Minister Guy Verhofstadt, a Liberal, who was a candidate for President of the European Commission. I presumed that Verhofstadt had asked this politician to smooth the way with the EPP. After discreetly inquiring among political friends, I did not comply with the request. All had said "No". We would nominate our own candidate if we won the European elections. However, not everyone shared this view. Our Benelux Prime Ministers, Jan-Peter Balkenende of the Netherlands and Jean-Claude Juncker of Luxembourg, unofficially supported Verhofstadt. But even they had a problem. When the word went around that German Chancellor Gerhard Schröder and French President Jacques Chirac were nominating Verhofstadt, it was abundantly clear that – like Helmut Kohl and François Mitterrand in 1994 – they had not reached agreement with the British. In the course of this evolving deadlock, the divisions over the war in Iraq had left deep scars: the Franco-German axis, supported by Belgium, was diametrically opposed to Britain.

In the Driver's Seat

On 1 May 2004, ten new Member States joined the European Union. During the Irish European Presidency, I invited our Prime Ministers to Dublin for an EPP Summit. Discussion focused on our candidate for President of the Commission. Juncker was against this basic assumption that we would put forward our own candidate. He contended that even if we remained the strongest party in Europe, we still did not have the best candidate. According to Juncker, the Commission President did not have to be appointed along political party lines. The Portuguese Prime Minister, José Manuel Durão Barroso, said he had a good candidate, his compatriot António Vittorino. As Vice-President of the European Commission, Vittorino had gained a good reputation at Justice and Home Affairs. But he was a Socialist and, therefore, could not receive our support. Perhaps Barroso preferred to have this powerful Social Democrat in Brussels rather than in Lisbon. Many participants at the EPP Summit, including myself, were shocked by this veiled support for non-EPP members. EPP Vice-President Peter Hintze expressed on behalf of the CDU his bewilderment at the lack of consensus. I was given the task of consulting with all the EPP Prime Ministers and working out who would be our best candidate. Hintze put it aptly: "Wilfried, you are once again in the driver's seat".

I wondered in all seriousness how I could resolve this complicated situation. In the course of the month of May, I spoke personally to every EPP Prime Minister. My first stop was Berlusconi and that was relatively easy. His position was perfectly clear: Verhofstadt, out of the question! This response was not exactly a surprise, in view of the poor relationship between the two. When I arrived in Athens, Kostas Karamanlis told me that Verhofstadt had called him up in person. And that proved to be the case not only in Athens. Everywhere I went Verhofstadt had already tried to make contact to argue for his candidacy and to point out that he had been nominated by Schröder and Chirac. Meanwhile, he was saying in Belgium that he would not stand. It would emerge that he was prepared to do almost anything to procure the coveted office of President of the Commission.

Verhofstadt's lobbying forced the EPP on the defensive. This was somewhat surprising, since we had, at least in theory, a consensus candidate, namely Jean-Claude Juncker. He was the ideal candidate. As the longest serving EU Prime Minister he enjoyed the trust of friends and foes. He was thoroughly familiar with the European dossiers and knew how to reach a compromise. But Juncker consistently declared that he would not stand, for a number of reasons, one of which was that the elections in Luxembourg coincided with the European elections. When I spoke to him on the telephone he stood his ground. Juncker would, nevertheless, have been an outstanding Commission President.

Tête-à-tête in Lisbon

After attending the congress of our Portuguese member party, the PSD, in Porto, I had a lunch appointment with Barroso at noon on Monday, 24 May in Lisbon. This meeting turned out to be crucial. During our *tête-à-tête* I spoke about the opposition aroused by Verhofstadt, but also about the prospect that we had no alternative, no candidate from the EPP. This would be a disaster because the Socialists had already taken important positions in the negotiations. Initially Barroso repeated what he had said in Dublin. He had a good Portuguese candidate and could not permit himself not to support him.

I said that he was a much better Portuguese candidate than Vittorino. Barroso had a solid reputation and, according to information gathered from the Council by former EPP Group Secretary-General Mario David, the French and the Germans thought exactly the same. At the end of the conversation, Barroso suddenly stiffened and said, "If you cannot find anyone else, I am prepared to stand".

My first meeting with Barroso dates back to the EPP Group study days in July 1994. Although he professes to be pro-American, I got to know him as a convinced European. His studies at the Centre Européen de Denis de Rougemont in Geneva almost certainly had something to do with this. In 1999, he became PSD President and Vice-President of the EPP. His election victory in 2002 was crucial. From then on his participation as Prime Minister at the European Council strengthened the position of the EPP considerably. In 2004, he had the courage to give up his premiership in favour of the Commission, "the most difficult job in the world", according to the title of a French TV documentary.

During a telephone conversation with the Austrian Chancellor, Wolfgang Schüssel, I mentioned the candidacy of Barroso. His reaction was positive. Schüssel himself would also have been a good President. But his candidacy had immediately aroused uncompromising opposition, principally from France. Within the EPP, the Germans hoped up to the last minute that Schröder and Chirac would change their minds in favour of Schüssel. Because he stood little chance, Schüssel himself did not press the issue for a moment, not even within the EPP.

Right up to the Tarmac

Late in the evening, one day after our latest European election victory, I received a telephone call from Verhofstadt. He said – as everyone now knew – that he had been nominated by Chirac and Schröder and also that he enjoyed great support from the CDU. I answered him, somewhat surprised: "Really? From whom in the CDU? I had a meeting with Angela Merkel last Monday in Berlin....

We agreed to support an EPP candidate". (I had told Angela that Barroso was prepared to stand.) "Really?" said Verhofstadt and added, "I will see her tomorrow at the airport when she comes to Brussels". Merkel was to take part in the EPP Summit on 16–17 June in Meise. They met on the tarmac. But I know that Merkel did not change her view when she met face-to-face with Verhofstadt.

During the EPP Summit on 16 June, all discussion centred on the new President of the Commission. No specific name had yet been mentioned at the time. Following close consultation with Angela Merkel and our group leader, Hans-Gert Pöttering, I sought contact with British Conservative Chris Patten. I asked him whether he was prepared to stand as the EPP candidate for the presidency. That evening I also had a telephone conversation with the leader of the British Conservatives, Michael Howard, who fully endorsed Patten's candidacy. This was a welcome bonus in the otherwise difficult relationship between the EPP and the British Conservatives. Perhaps, just as in 1992, when we started our collaboration with the British Conservatives, Patten could act as a bridge. The choice of Patten was important: he was one of us. As European Commissioner he was a great help to the EPP – on top of that, he was in effect a Christian Democrat who completely endorsed our way of thought. He was universally respected for his diplomatic skills and for his impressive work at the Commission on external relations. He was British and would be able to rely on the support of Prime Minister Tony Blair. He was pro-European and this might give a positive turn to the British Conservative Party.

Three-Cushion Billiards

On Wednesday morning – 17 June 2004 – we met again to nominate a candidate. There was a huge turnout from the press. Everyone assumed that after our victory, we would have a clear favourite. Before the meeting, I spoke to each of our Prime Ministers separately to let them know about Patten's candidacy. Greek Prime Minister Kostas Karamanlis hesitated because a British candidate was a very sensitive issue in view of the Cyprus question. I could also detect doubt in Barroso's mind. I confided in him: "Surely you can see that we are playing 'three cushions' here. It is perfectly possible that the ball going in Patten's direction will ultimately end up coming to you". Berlusconi, the most senior Prime Minister in terms of age, was prepared to argue for Patten's candidacy with Tony Blair. According to the Treaty of Nice, the proposed President of the Commission would have to secure a qualified majority of votes in the European Council. This majority was not attainable without the support of Blair and the EPP Prime Ministers. At the same time, together we formed a block against the proposal by Schröder and Chirac and, therefore, against the candidacy of Verhofstadt.

Pöttering sat next to me at the meeting and prodded me to suggest a name. I then proposed Patten. Barroso stuck to the candidacy of Vittorino. Juncker continued to defend Verhofstadt. Berlusconi stated frankly: "I need to know what I have to do at the European Council". In response to my suggestion, everyone finally agreed that Berlusconi would announce Patten as the EPP's choice and that he would ask for Blair's support. I had often said that the outcome of the European elections should be the deciding factor in choosing the new President of the Commission. This objective was now within our reach. Yet I stayed clear of the press. Pöttering publicly stated his opposition to Verhofstadt, thus keeping me out of the line of fire.

At the European Council, it became immediately apparent that there was no consensus in favour of Verhofstadt. Berlusconi had Blair's agreement to support Patten, not Verhofstadt. Chirac and Schröder spent another night trying to persuade their colleagues. But the minority blocking Verhofstadt held strong. Finally, Chirac made it clear to Verhofstadt that he no longer stood a chance. At the same time, however, he also opposed Patten. Maybe the idea of the Commission being led by a Briton – especially after Blair's opposition to his candidate, Verhofstadt – was too much for Chirac to swallow. When I decided to ask Patten to stand, I knew that his candidacy might provoke opposition, and my conscience was uneasy when I approached him. Patten was aware that the effort might not succeed: the opposition came from Paris. But this left the field clear for another candidate, and Barroso was ultimately appointed. In the end, the "three cushions" game produced a positive result for the EPP. If the candidacy of Barroso had also run aground, we would then have nominated, from the EPP, Michel Barnier, former European Commissioner and at the time French Minister of Foreign Affairs. We had the full support of the French Prime Minister, Jean-Pierre Raffarin.

Among Believers

I kept my distance from what was going on in and around the European Council. I was already relaxing on the coast of Belgium when Chirac finally endorsed the candidacy of Barroso. Angela Merkel called me to say, "We both have a reason to be happy. We have succeeded in our mission". Undoubtedly, her intensive communications with Blair contributed greatly to the final outcome. The nomination of Barroso at the special European Council of 29 June 2004 was a victory not just for the EPP but also for those who never gave up believing that it would be possible to capitalise on our victory and exercise our democratic right to appoint an EPP candidate to head the EU. Angela Merkel was one of them. Without her I would never have succeeded in pushing Barroso to the forefront. That too was a masterful example of three-cushion billiards! These are the moments when close personal ties are forged.

I got to know Angela Merkel when she was a minister in Kohl's governments. The evening before a CDU Congress I was sitting at a table with Kohl and Merkel when Helmut said to her, "Angela, du sollst dem Vorzitsender einmal einladen in Mecklenburg-Vorpommern" (Angela, you have to invite the President to visit Mecklenburg-Vorpommern), the state where she lived. And she did, in fact. But, unfortunately, due to my heavy agenda I could not take her up on the invitation. When she became CDU President in 1999, she immediately came to see me in Brussels. We have always remained in very close contact since. She is one of the government leaders I can speak to on her mobile telephone or send a text message to at any time. Kohl used to call her *das Mädchen* (little girl) in those days, but even at a very early stage she showed signs of great political intelligence. She defends her convictions, supports the EPP to the hilt and stands on her own two feet in international politics. In the tradition of Kohl, she took on the leadership of a new generation of EPP leaders. As Chancellor she was responsible for the great success of the German Presidency of the EU in the first half of 2007. She saved European finances from disaster by forging a consensus on the 2007–2013 budget. Under her leadership, the European Union has become intensely engaged in the battle against climate change. And in June 2007, she successfully reached an agreement on a new Reform Treaty that was later signed in Lisbon in December 2007 and has breathed new life into the European Constitution, which had been presumed dead. In a word: she has set the stage for a re-launch of the European integration process.

Believers are rare, for the difference between victory and defeat is wafer-thin. It was unusual, and a first in every way, for the Commission President to be appointed because of the direct intervention of a European political party. We needed therefore to overcome the division among our own Prime Ministers. As Party president I also had to be prepared to work on the sidelines for a solution – away from the microphones and television cameras. Timing was also crucial. If we had nominated Barroso as our candidate straightaway, he might have confronted Franco-German opposition, because he had organised the Azores Summit with the supporters of the attack on Iraq.

Whether this scenario could ever be repeated is doubtful. Will our EPP member parties and their European foot soldiers, the MEPs, be allowed to express democratically their opinion on who should lead the EU? Or will this decision be once more monopolised by the heads of state and government, regardless of the result of the European elections? Will our political opponents ever again grant us such success?

A Changed Field of Influence

The installation of the Barroso Commission was only the beginning. After Barroso's election, we had to resolve various problems related to the approval of the new Commission by the European Parliament. I was closely involved in the crisis over the Italian Commissioner-designate, Rocco Buttiglione. He had come under fire in Parliament because of controversial comments about homosexuals and the role of women. At Barroso's request, I also mediated between EPP-ED Group chair Pöttering and Fidesz President and former Hungarian Prime Minister Viktor Orbán. Orbán, who was in opposition in his country, needed to be persuaded to abandon his objection to the contested candidate of the Hungarian government, László Kovács, in exchange for a reshuffle of portfolios. Our group was prepared to overlook the *casus belli* against Kovács and to tolerate him. This formed part of a compromise in which the controversial Buttiglione was replaced by Italian Foreign Minister Franco Frattini.

The election of the Barroso Commission illustrated the shift in the field of influence between the institutions. The European Parliament was no longer the "toothless tiger" when it came to decision-making. It showed its teeth against the European Council and against the Commission. But this political battle was also animated by differences of opinion between the political groups. The choice of Barroso and the crisis over Buttiglione proved the importance of timing. Both Parliament and Commission wanted to seize the initiative at the beginning of the new legislative term. The fact that the five-year terms of the two institutions coincided also increased the intensity of their animosity. This was just one of the many signs that institutional relations were evolving into a tense struggle between the executive branch, the Commission, and the legislative branch, the Parliament, as is exactly the case in national politics with government and parliament.

A Strong Position Confirmed

The fact that the EPP was in a strong position was already apparent at our congress in Brussels on 4–5 February 2004. We were addressed by an impressive array of government leaders. This has not always been the case in the history of the EPP. Because José María Aznar was no longer standing in the Spanish elections in March, it was his last congress as leader of the Partido Popular. A new generation was standing in the wings or was already in office: Angela Merkel in Germany, Edmund Stoiber in Bavaria, Jan-Peter Balkenende in the Netherlands, Jean-Claude Juncker in Luxembourg, Wolfgang Schüssel in Austria, Kostas Karamanlis in Greece, José Manuel Barroso in Portugal, Silvio Berlusconi in Italy, Jean-Pierre Raffarin in France and so on. The Action Programme for 2004–2009 aroused little controversy. All of the attention was focused on the enlargement of the EU on 1 May. In a sense the congress came too early, because the campaign for European elections had not yet started, not by a long shot.

The European elections in June 2004 confirmed our strong position. With 268 MEPs, we were again the largest group by a wide margin in the new, enlarged Parliament. The Socialist Group had 200 members. Forty per cent of the MEPs from the new Member States joined our group. This meant that our group was the only one in Parliament that had representatives from all Member States. This victory was largely due to the good performance of our new member parties in the countries that had just joined. Years of pioneering democracy-building work in Central and Eastern Europe had not been in vain. Now we could reap the rewards.

Primarily National Ambitions

In the run-up to the European elections, there was a dramatic development. It was generally known that the French UDF and the Italian PPI were not very keen on the enlargement of the EPP. Other parties, such as the Flemish Christian Democrats, also often had difficulty with this strategy. But when our group decided to revise its internal rules of procedure to please the British Conservatives, the French centrists and the left-oriented Christian Democrats of Italy cried foul. They turned it into a matter of principle. When they lost the argument, they abandoned our political family. After the European elections they joined Bayrou's European Democratic Party and, therefore, the Liberal Group in the European Parliament.

In actual fact the leaders of our group had given them an excellent excuse to do what they had already planned to do for a long time: to pursue their own course at the European level. The UDF and PPI found it difficult to stomach the fact that after the accession of the UMP and Forza Italia they were dominant within the Party. Another reason for their ongoing hostility was that they had chosen opposition, whereas the UMP and Forza Italia had government responsibilities. They did not do what other member parties had in fact been able to do: form a single centre-right delegation and combine their powers in the European Parliament. To paraphrase von Clausewitz, they saw European politics as the extension of national politics by other means.

UDF President François Bayrou, in defiance of the majority of his supporters, had refused to join the UMP. It is abundantly clear that he had been waiting for the right moment to promote his own ambitions, which he ultimately did in the French presidential elections of 2007. Above all, Bayrou wanted to remain *présidentiable* as a centrist candidate and to offer an alternative to centre-right voters. This was the real reason for his departure. But we also know that his presidential dream is now a distant memory and that even his most trusted friends have abandoned him. Because of his personal ambitions, the damage has now been done. The CDS – the successor of Robert Schuman's party, the Mouvement Républicain Populaire, precursor of the UDF and one of EPP's founding parties – no longer exists.

The same applies to Romano Prodi. In the Italian parliamentary elections in April 2006, he put himself forward as an alternative to Berlusconi. He needed the support of the PPI, one of the main successors to the Democrazia Cristiana. He persuaded the PPI to join the new European Democratic Party. The fact that he was still President of the Commission during all these machinations was apparently not a problem for him or his supporters and secret sympathisers. Subsequently, the left-oriented Christian Democrats of the PPI joined Margherita and afterwards, together with the post-communists of the Mayor of Rome, Walter Veltroni, formed the Italian Partito Democratico (PD). This party is the de facto merger of the Democrats of the left, distant heirs to the communist PCI, with Margherita, a combination of various groups from the Democrazia Cristiana. At the close of the PD's founding congress, the Christian Democrats sang the "Internationale" along with their Socialist comrades ....

The Hard Core

A party like the EPP, which faces such an array of challenges, can be ready for the fray only if it can fall back on the expertise and devotion of its staff. It is, therefore, not surprising that our party headquarters in Brussels requires a high degree of continuity and has often been staffed by Belgians, especially in the early years. The simple fact of proximity, availability and permanent accessibility played a decisive role. More recently, the Party has attracted staff from many different corners of Europe who made the commitment to relocate to Brussels and dedicate themselves to the EPP. The fact that these hard-core "political soldiers" live in Brussels undeniably contributes to the power of the EPP.

European political parties like the EPP must be able to operate in a stable environment. Their official status, and the funding from the European Parliament associated with this status, has greatly contributed to this. I believe I can claim the political responsibility for this reform. The lack of rules was long a sore spot. When I became President of the EPP in 1990, I immediately spoke to Commission President Jacques Delors. I called my contacts at the European Council, put the matter on the agenda of the EPP Summit and convened my fellow Prime Ministers. The other European parties had an even greater need of legal status than the EPP. Their employees were on the payroll of their parliamentary groups and they were like squatters taking refuge in Parliament. From the very beginning, we had our headquarters outside the Parliament buildings.

At the European Council of Maastricht in December 1991, I proposed to my fellow government leaders that we should enact a statute to legitimise European political parties. My proposal was based on a clause in the German Constitution. The article which, after a great deal of effort, became part of the treaty proved to be unsatisfactory because it was purely declarative. There followed years of intense discussion about its legal basis and how to implement it. There was a struggle between politicians and diplomats, especially the permanent representatives of the Member States. When I was elected to the European Parliament in 1994, I tried to force a breakthrough together with the other group leaders. All kinds of formulae made the rounds, but none of them proved to be the answer until we asked the legal service for advice. The Parliament's lawyers informed us that there was no legal basis for enacting the rules on the status and funding of European parties. After all, the EU has only those powers that are allocated to it. I felt betrayed!

The legal basis was ultimately provided by the Treaty of Nice, specifically Article 191, paragraph two: "The Council establishes, according to the procedure of Article 251 (co-decision), the statute of the European political parties and particularly the rules concerning their funding". However, recognition of European political parties still had a long way to go because some governments remained opposed to the regulation. An initial attempt sponsored by the German Ursula Schleicher, a member of the EPP Group, foundered in the course of 2001. A second attempt by Parliament, sponsored by the German Socialist member Jo Leinen and the Greek EPP member Giorgos Dimitrakopoulos, was finally successful. At the end of 2003 the Council approved the regulation with a qualified majority.

Today, ten European political parties have statutes. They receive an allocation from the European Parliament calculated on the basis of the same minimum sum for each and an additional amount *pro rata* based on the strength of the Group. In accordance with the new rules, we have reconstituted the EPP from a natural association into a legal body, specifically an International Non-profit Association, a legal arrangement in Belgian legislation. The system is now fully transparent and legal; abuse or improper use of party subsidies is absolutely forbidden. This may seem like a Pyrrhic victory, yet it constitutes an enormous achievement after a long and difficult journey through the European institutions.

Our work is not over, however, since the statute is not complete. But it would have been foolish to press for more and there was the ever-present danger of upsetting the apple cart. It would have been impossible to ask for a complete package, which would include the incorporation of a uniform organisation for European elections, a statute for the MEPs and so on, in a single step. But it was certainly possible to ask for financial support. The sponsors took my worries and warnings to heart. With this compromise, I too was able to put my stamp on the recognition of European political parties at the end of this long process.

The new regulation, nevertheless, threatens to become a victim of its own success. Since so many parties have submitted applications and received funds, the European party landscape is threatening to become a kind of patchwork quilt. As the number of parties that pass the minimum threshold increases, the relative proportion for the larger parties is reduced, since the total sum remains constant. Moreover, even Eurosceptic parties that fought against this regulation receive financing from Parliament. The opponents are, as it were, being rewarded for their opposition!

There is a great risk that European parties will end up living in a glass bubble and will neglect the need to cultivate their own organisations. If there is no real, permanent and motivated political leadership on board, the parties will very soon evolve into organisations that spend most of their funds on maintaining their staff, for example. Political actions then degenerate into pure symbolism. This means that the ultimate goals of the party are lost sight of. Attention is paid chiefly to material and staff concerns. In my opinion, a party needs to be permanently on the move. It cannot become an interest group for those who have the privilege of working in and around European institutions. Only too often, exaggerated importance is attached to financial privileges, personal ambitions and the length of titles on business cards.

I also wanted to see the EPP develop into a party of ideas. To win the battle of European integration, it is not so much new or supplementary structures that are necessary. The European parties need to be staffed by people who want to play a role in "the European concert", as Jacques Delors so aptly described it. This can be done by bringing together ideas and expertise from the existing research departments of the parties and other institutions. Some parties have outstanding think tanks, such as the Konrad Adenauer Stiftung, which is linked to the CDU, the Hanns Seidel Stiftung, linked to the CSU, or the FAES in Spain, of which José María Aznar is President. Within the EPP-ED Group, the British Conservative MEP James Elles worked tirelessly to set up the European Ideas Network, which organises an annual Summer University and an Ideas Fair for politicians, academics and journalists.

The EPP once again took the lead in establishing a fresh, innovative European political think tank. This became possible following the revision of the regulation on European parties, which, after another strenuous roller-coaster ride with the institutions, was finally completed in December 2007. The regulation allows European parties to set up "European political foundations". It is actually astonishing to think that the EPP, which has existed for over thirty years, has only now been allowed to establish its own research centre. Our new Centre for European Studies has started to engage the research departments and foundations of our member parties in an effort to pool all of our resources in our common European think tank. The time has come for our think tank to introduce ideas and pump oxygen into our political family, European institutions, and Europe and its citizens. The start has been modest, but this does not distract us from the importance of the task. Here too "bureaucratic temptation" may raise its ugly head. It will, therefore, be necessary to work hard to realise our ambitions.

The Party of Continuity

The crucial importance of the EPP Summit for the development of the Party has often been evident. Consultation between our government and party leaders dates from the beginning of the 1980s. At the end of that decade this conference of party and government leaders became the political trademark of the Party. In the run-up to the European Summit at Maastricht in 1991, it reached a high point of both frequency and efficiency.

For several years now, the European Council has met permanently at the Justus Lipsius Council Building in Brussels. As Prime Minister, I was able to organise the EPP Summit meetings at Val Duchesse, a castle close to the European quarter in Brussels. We later moved to the buildings of the European Parliament. Thereafter we convened at Bouchout Castle, situated in the grounds of the Meise Botanic Gardens near Brussels.

One or more controversial issues from the agenda of the European Council are selected as topics for our discussion at the EPP Summit. We try to reach a common viewpoint with our government leaders. I also make sure we discuss programmatic evolution, important decisions on new EPP members and how politically up-to-date the member parties are. Things are done informally and extensive paperwork is not allowed. The crucial thing is the timing of the meeting. It is actually a problem that we cannot meet in the morning or the day before the European Council. Helmut Kohl, in his day, was absolutely convinced that in order to be able to give proper consideration to the decisions, the preparatory discussion ought to be held a week or two before the European Summit. This was done systematically right to the end of his chancellorship. He set the tone for all the discussions. Without his efforts we would never have been successful. And because he was so strongly committed, the others were obliged to follow him. Angela Merkel resumed and continued this tradition, especially during the German EU Presidency in 2007.

This gathering gradually became larger. Experience had taught us that it was also necessary to invite the EPP opposition leaders, because at some point they will take on government responsibility. Furthermore, we wanted to involve government leaders of the EU candidate Member States and of the countries from the Council of Europe in our political project. The EPP Summit has, therefore, developed into a major event in which all the prominent leaders from our political family participate.

In all these developments the support of the CDU/CSU has been indispensable. The German Christian Democrats are in many respects the most important member party of the EPP. At the time of the founding of the EPP, through the efforts of Hans-August Lücker and in spite of the problems surrounding the EDU, they already played a decisive role. This commitment and the inspiring role played by the German Christian Democrats have never diminished. We owe this to the strong leadership of Helmut Kohl and Angela Merkel. The CDU/CSU is a beacon of continuity in the life of the EPP. This leadership was never overbearing or arrogant but, on the contrary, marked by a sense of co-determination and plurality. Very striking is the room the German Christian Democrats allow – grant, as it were – to the member parties from smaller Member States. For instance, it is perfectly natural to them that all the EPP presidents so far have come from one of the Benelux parties.

I have attended almost every *Parteitag* of the CDU for thirty-five years. The CDU has strong central management. The leadership is always in the hands of the party president, who is the federal Chancellor when the party is in power and, when it is in opposition, is group leader in the Bundestag and the CDU Presidium that meets every Monday in Berlin. Yet the party cannot be understood without knowledge and understanding of the actions of the *Länder*. So this is certainly no Latin American presidential system. The internal federalism is genuine. The party is also in a strong position intrinsically. Decisions taken at the CDU Congress are thoroughly discussed in all party departments.

The congress in Hannover on 3–4 December 2007 was dedicated to the new *Grundsatzprogramm* (core programme). The party meeting had been prepared months beforehand by the Secretary-General, Ronald Pofalla. True to tradition, it opened with an ecumenical service and homilies by a Catholic and a Protestant bishop. Young men and women played an important part in the decision-making. They interpreted the ideas and objectives of their generation in continuity with the great tradition of their party. With *die Mitte,* a term that is far more precise and more apt than the term "centre-right" or "Conservative", they proposed Christian Democratic policies for Germany and Europe that are bound to inspire all EPP member parties.

Political Leadership

The German Christian Democrats form the backbone of the EPP and the European Union. That was invariably the case in the time of Helmut Kohl and turned out to be so again during the German EU Presidency in the first half of 2007. Angela Merkel performed her challenging and sensitive task magnificently. Thanks to her personal commitment, dedication and efforts, considerable progress has been made in three important domains: the EU multi-year budget, the fight against global warming and, above all, the salvaging of the European Constitution. In this case a breakthrough was particular crucial. Merkel took this hugely difficult task to heart. She succeeded in forcing the European Council of 21–22 June 2007 to reach a consensus on a new Treaty. I am especially pleased that – through the EPP Summits of Helsinki and Meise on 19 October 2006 and 21 June 2007 – the party contributed to this by first agreeing unanimously within its own political family how the institutions should be reformed.

Our congress in Rome in March 2006 on the eve of the Italian parliamentary elections came too early for a definite decision, because the change had not fully taken place in people's minds. Yet the concluding document, "For a Europe of the citizens: Priorities for a Better Future", already contains all the essential elements that have since been reworked to form a consensus. Carried out during a reflection period after the negative referenda in France and the Netherlands, this was not a superfluous exercise. As all observers of Europe knew, people were waiting for the Dutch parliamentary elections and the election of Nicolas Sarkozy as French President on 6 May 2007. Because Germany's vote weighed so heavily, some governments, such as those of Poland and the Czech Republic, could not back out of the consensus. The most difficult exercise still remains the ratification of the Lisbon Treaty in a few remaining Member States. The rejection of the Treaty by the Irish in the June 2008 referendum has created a new setback for the EU. I hope that the leadership in Ireland will find the courage to resolve this impasse for the good of their country and for the good of Europe. We need to put an end to our esoteric discussions about institutions and start dealing with the real needs of our citizens. The Lisbon Treaty may not be sufficient, therefore, but it is necessary. Rather than being the end, it needs to be the new beginning.

The crisis that preceded the Treaty of Lisbon led to a thorough shake-up of relationships within the EU. In France the referendum was a clear vote of sanction against Chirac, spurred on by the extreme left and the extreme right as well as by the absence of the internally divided Socialists. The UMP and, more specifically, its then President Nicolas Sarkozy, prepared the way out of the impasse. Already in his address on 6 September 2006 at the Solvay Library in Brussels, at the invitation of Friends of Europe, he proposed a concept that would later become the basis for a solution. Via a *traité simplifié*, he built a bridge between those who were sticking to the Constitution and those who had declared it dead. Thanks to his intervention, Angela Merkel was able to reach a consensus on a new treaty which reinvigorated the European Union.

The positive referenda in Spain and Luxembourg are the fruit of political leadership. They were not a foregone conclusion. In Spain a responsible opposition party resisted the temptation to reach for short-term political advantage. It refused to follow a Eurosceptic path. A great deal of credit is due to the leadership of the Partido Popular and Mariano Rajoy which were, at the time, in a bitter struggle with the ruling Socialists.

We owe the Luxembourg victory to the personal efforts of Prime Minister Jean-Claude Juncker. His commitment to Europe is deeply rooted in the traditions of the Luxembourg Christian Democrats. Since it was founded, the CSV has played a very active role in the EPP, which is evident from the involvement of its leading figures. Jean-Claude, Prime Minister of Luxembourg since 1995, is a strong voice at EPP summits and in the European Council. His prestige is also the result of his expertise and efforts as chair of the Euro Group. His predecessor, Jacques Santer, served with distinction as President of the EPP and the European Commission in difficult times. Santer's two illustrious predecessors, Joseph Bech and Pierre Werner, played a historic role in the genesis of European integration, the latter with the Werner Report, which formed the basis of the euro and the Economic and Monetary Union (EMU). It is remarkable how a small country such as Luxembourg – until 2004 the smallest Member State in the Union – has, through political leadership, played such an important part in the unification of Europe.

A Cultural Homeland

The same can be said of the Netherlands. Even though at that time the Dutch Christian Democrats were still three separate parties, within the EPP they were with us from the outset. Frans Andriessen represented the KVP, Hans de Boer the ARP and Roelof Kruisinga the CHU. The EPP has reaped many benefits from their merger to form the Christian Democratic Appeal (CDA) under the leadership of Piet Steenkamp. Piet Bukman was EPP President from 1985 to 1987. Jos van Gennip, via the Scientific Institute, drafted the Basic Programme of Athens. In the 1990s, Wim van Velzen played a prominent role in the EPP Group, but his great, historic merit lies in his work within the Party: first as the architect of the merger of the EUCD with the EPP and subsequently as the person responsible for new member parties and our irrepressible growth. His successor and a rising star, Camiel Eurlings, has meanwhile exchanged the European scene for national politics but that does not mean that his role is over. Corien Wortmann-Kool has taken over his duties with particular talent and dedication.

My mother tongue is spoken in the Netherlands. The Dutch language has played a crucial role in my personal and political life. As a young student I was fascinated by *Max Havelaar*, a masterpiece of Dutch literature. The book is an indictment of the treatment of the Javanese in the Dutch colony of the East Indies; it was written in the nineteenth century by Multatuli, the pseudonym of Eduard Douwes Dekker, who had acted as Assistant Resident in the district of Lebak. I have often meditated on his battle against corruption and his address to the Chiefs of Lebak which begins thus:

> We are all in the service of the King of the Netherlands. But He who is just and wants us to do our duty is a long way from here. Thirty times a thousand times a thousand souls, more than that, even, are bound to obey his commands, but he cannot be near all those who depend on his will. (…) And I, who yesterday took almighty god as my witness that I would be just and merciful, that I would do right without fear and without hate, that I will be "a good Assistant Resident"… I too wish to do what is my duty. Chiefs of Lebak! That is what we all wish!

He then speaks about a dead man who was just and merciful:

> He was good and just. He administered justice and did not drive the complainant from his door. He listened patiently to anyone who came to him

and reported what he had heard. And he helped anyone who could not drive his plough through the ground, because his buffalo had been stolen from his stall, to look for the buffalo. And where the daughter had been snatched from the mother's house, he sought the thief and brought the daughter back. And he did not withhold wages from those who had toiled and he did not take the fruit of those who had planted the tree, and he did not dress himself in the clothes that were supposed to cover others, nor did he feed on the food that belonged to the poor. So, in the villages they will say, "Allah is great, Allah has taken him to himself. His will be done … a good person has died". (Gutenberg Ebook of Max Havelaar, p 63–64, our translation)

I have an emotional bond with the Netherlands and feel both a spiritual and a physical closeness to this country. I call it my cultural homeland. I also feel a great intimacy with the landscape because of its similarities to the area where I was born.

As a young man I explored the Zeeland (Zealandic province) by bicycle; today, many years later, I am doing this again in Veluwe, the northern part of the province of Gelderland (Guelders).

My first political contacts there go back to my student days. My Dutch peer Erik Jurgens and I jointly organised Flemish–Dutch student demonstrations. He too later chose the political path, first with the progressive Christian Political Party radicals and afterwards with the social-democratic Labour Party. During my period in office I established close collaboration with my colleagues, Prime Ministers Dries van Agt and Ruud Lubbers. I was also a frequent guest of the Dutch media during that period.

In recent years, cracks have appeared in my image of the Netherlands. The crisis following the rise and murder of Pim Fortuyn is not yet over. I was also convinced that the referendum on the European Constitution would fail. Euroscepticism has been fostered for years, particularly by the then Minister of Finance Gerrit Zalm, with his comments that the Dutch were paying too much to Europe. First the administration and the politicians changed track, and so ultimately did public opinion. Added to this came the excessive importance attached to national symbols and the fuss about organising a so-called non-binding referendum, leaving the Parliament of the Netherlands aside completely and powerless to ratify the Constitution. Seeing this lack of continuity has left me dumbfounded.

Today a change is taking place. The big question is whether the Dutch political class will still be capable of following the European traditions of its predecessors. All eyes are fixed, in the first place, on the Prime Minister, Jan Peter Balkenende, to restore this continuity. I am convinced that he will succeed.

Chapter X

Pushing Forward in Europe and Beyond

Immediately after my resignation as Prime Minister on 6 March 1992, I moved to the headquarters of the EPP in Brussels. I dedicated myself entirely to the organisation of the Party. Over the next two years – from March 1992 to April 1994 – I undertook ninety-one trips, a physically exhausting task, to reorganise and represent Christian Democracy. I led numerous fact-finding missions in search of partners. A journalist described me then as "the traveller for democracy".

Christian Democracy remained unknown for a long time in Central and Eastern Europe, although Christian-Social parties represented a considerable part of the political landscape in Central Europe in the nineteenth and twentieth century. The origin of these parties in Czechoslovakia and Slovenia dates back to the 1890s. In Hungary the first Christian-Social political formations appear in 1885, and in 1894 the Christian People's Party was founded. As of 1898 and 1901 the Christian-Social–inspired Poles, then under Austrian and German rule, assembled as political parties.

The EUCD's Final Role

The more the EPP grew in importance, the more the European Union of Christian Democrats (EUCD) declined. The EUCD was a loose Christian Democratic organisation that preceded the EPP but, oddly enough, it continued to function in parallel to the EPP. Already in the 1980s there were plans to absorb this organisation into the EPP. Another centre-right organisation, the European Democratic Union (EDU), also functioned in parallel to the EPP. Needless to say, the existence of these three organisations – which sometimes worked against each other – gave the outside world a convoluted impression of the European centre-right. Our political family looked fragmented, and not a single outsider or observer could understand why we organised ourselves in such a labyrinthine manner.

Yet after the fall of the Berlin Wall in 1989 and the revolutions in Central and Eastern Europe, the EUCD had a new reason for existing. It could serve as a waiting room and training school for parties in those countries destined to later join the European Union. In anticipation, the EUCD offered an intermediate forum where they had a chance to assemble and meet; at the same time, the EPP avoided all pressure to expand quickly with parties that did not yet meet its standards. When in January 1993 I succeeded the Italian Emilio Colombo as the leader of the EUCD, the leaderships of the EPP and the EUCD were *de facto* transformed into a personal union. I continued to work in the headquarters of the EPP, with the position of EUCD chair being an additional task for me until October 1996.

I came to realise that besides East and West there was another distinction and that a virtual border runs through the European cultural heritage, from Saint Petersburg to Zagreb. This separation dates back to the Great Schism of 1054 between the Eastern Orthodox Churches and the Roman Catholic Church, which created two different societies: one along the mould of Western Christianity and the other marked by the Eastern Orthodox Christian tradition. This socio-cultural division remains as much a fact today as in the previous centuries; a division between, on the one hand, Western and Central Europe and, on the other hand, Eastern Europe. Since the fall of the Iron Curtain, it has been more difficult for pluralistic democracy to develop east of this dividing line; economic reforms are slow and civil society has a hard time manifesting itself against political and military power. This distinction is more deeply rooted than the artificial separation between East and West that resulted after the Second World War.

With my election in July 1994 as chair of the EPP Group in the European Parliament, I became responsible for a new and heavy task. I was entrusted with three simultaneous political leadership mandates: President of the EPP, chair of the EUCD, and chair of the EPP Group in the European Parliament. This heavy combination was no longer endurable for me. I insisted that the veteran Dutch politician Wim van Velzen succeed me as chair of the EUCD. Helmut Kohl wanted an immediate merger of EPP and EUCD; in this way, according to him, my personal conflict of mandates would be resolved. Kohl also saw no value in a transitional arrangement with van Velzen. He feared that this would lead to the indefinite continuation of the EUCD.

It took a lot of effort to obtain Kohl's agreement that Wim van Velzen become the EUCD chair. I finally convinced him that under van Velzen's leadership the merger of EUCD and EPP would finally take place. It did take some time. On 2 February 1996, Helmut Kohl received an Honorary Doctorate from the Catholic University of Louvain. After the ceremony, I accompanied him in his car to Laeken where he would be received by King Albert. During the car ride I had a final conversation with him where, ultimately, he gave his agreement for van Velzen.

Wim van Velzen played an important role in our efforts to promote democracy. On 6 October 1991, the EPP founded an academy in Budapest for the promotion of democracy in Central and Eastern Europe, the Robert Schuman Institute. Van Velzen chaired this Institute for many years and produced impressive results. Van Velzen also chaired the EPP's Working Group for new EPP member parties. All those who were involved in our painstaking efforts know that van Velzen played a pivotal role in anchoring our political values in the transition countries of Central and Eastern Europe.

The Hungarian Struggle

In Germany and Austria, Central Europe refers to the territory covered by Germany, Austria, Switzerland, Liechtenstein, Poland, the Czech Republic, Slovakia, Hungary and Slovenia, and often includes the regions that were part of Austria-Hungary. The essayist and novelist Claudio Magris, from Italian-Slovenian Trieste, wrote about the inscrutable *Mitteleuropa*, a masterpiece fed by the Danube.

My first participation in a political demonstration was in 1956 on behalf of the Hungarian uprising. I was studying at the University in Louvain and, like many of my fellow students, I was deeply outraged at the Soviets for crushing it. We demonstrated in Brussels against the Soviets until late at night. Today, more than fifty years later, our friends in Fidesz would like to erect a monument in the Leopold Park in Brussels commemorating their heroic struggle for freedom and democracy.

The Hungarian Revolution reminds me of János Kádár, who in 1956 brought about the downfall of Imre Nagy, leader of the revolution. Kádár became the new leader of the Communists and I met him as Prime Minister in 1984 during an official visit. My Hungarian friends informed me later that he was plagued by nightmares because he was responsible for the secret summary executions of young demonstrators, who after the revolt were first imprisoned for many months. At the end of our meeting he requested an audience with King Baudouin. His wish was granted one year later. At the end of his visit I escorted him; by then he was already an old and worn-out man. When the communist regime became fragile he was sent into retirement on grounds of poor health. Kádár stayed in power until 1988 as a loyal vassal of the Soviets.

The origins of Christian Democracy in Hungary date back to the end of the nineteenth century. However, in 1920 the Christian People's Party (founded in 1894) became a conservative party and renamed itself Christian Union. It stayed in power until the war and identified itself with a regime that became increasingly powerful, reactionary and authoritarian.

In 1940, Christian reformers who left the governing Christian Union formed the People's Social Movement. Although the attempt failed, it became the basis of a new political formation, the Democratic People's Party. On 31 August 1947 it obtained 16.4% of the votes and thus became the second-largest party in Parliament.

As the strongest opposition party it resisted authoritarian rule and defended fundamental rights and freedom of religion until it was dissolved in January 1949 by the communists.

On 1 May 1989 it was refounded as the Christian Democratic People's Party (KDNP), heir and successor of the Democratic People's Party. The KDNP was a modest but above all a new political power. In spring of 1990 it was elected to Parliament in the first free elections in fourty three years. József Antall, leader of the Hungarian Democratic Forum (MDF), became Prime Minister on 23 May, governing as leader of a coalition between the MDF, KDNP and the Independent Smallholders' Party (FKGP), who together obtained 59.5% of the votes. In the EUCD, the MDF and KDNP formed the "Hungarian team".

On the invitation of József Antall the congress of the Christian Democratic International (CDI) was held in Budapest at the end of June of that year. I remember this as a real political highlight. After the fall of the Berlin Wall new democratic governments were being elected throughout Central Europe. Many eyes were drawn with hope to the leaders that were present. The atmosphere was at times one of elation. After the official dinner Helmut Kohl insisted that his colleagues, led by József Antall, walk over the chain bridge from Pest to Buda to a famous restaurant to taste his revered salami ... We enjoyed this until late at night whilst strengthening our friendship. Unfortunately, this kind of political bonding was seldom repeated.

Sadly, József Antall was unable to complete his first term as Prime Minister. Stricken by cancer, he passed away on 12 December 1993 and with him the leadership of his party disappeared. His successor, Peter Boross, lost the elections in 1994 and the Christian Democratic parties went into opposition. Their reawakening would occur four years later with Viktor Orbán.

Fidesz – Hungarian Civic Union

Orbán was a founding member of Fidesz (Alliance of Young Democrats), created in 1988 by young democrats, mainly students, who were persecuted by the Communist Party and had to meet in small clandestine groups. The movement became a major force in many areas of modern Hungarian history, engaging itself on every level in the development of a democratic system, its members being active as guardians of fundamental human rights. There was an upper age limit to membership of thirty five years but this was abolished in 1993.

On 16 June 1989, Orbán gave a speech at Heroes' Square on the occasion of the reburial of Imre Nagy and other national martyrs, in which he demanded free elections and the withdrawal of the Soviet troops. The speech brought him wide national and political acclaim. In the summer of 1989 he took part in the

Opposition Roundtable negotiations, and in the 1998 elections he was candidate for Prime Minister. Under the direction of Viktor Orbán, Fidesz was transformed from a radical student movement into a moderate, centre-right people's party. In September 1992 he was elected vice-chair of the Liberal International. However, in parallel with its transformation into a centre-right people's party, Fidesz left the Liberal International and became in November 2000 an associate member of the EPP. Since then, Orbán has been working hard to strengthen the centre-right alliance in Hungary. During the EPP Congress in Estoril in October 2002, he was elected EPP Vice-President.

In 1998 Viktor Orbán formed a successful coalition government with Fidesz and two junior partners, the MDF and FKGP. The leader of the MDF, Ibolya Dávid, proved to be a remarkable Minister of Justice and enjoyed great popularity. But after a heavily polarised election campaign the government lost its majority in 2002. All eyes were on a comeback in 2006. Fidesz formed an alliance with KDNP and added "Hungarian Civic Union" to its name. However, relations with MDF turned sour. Dávid accused Orbán of trying to absorb her party. This was a bad omen and, not surprisingly, the two parties had a bitter quarrel in the middle of the two rounds of the 2006 elections. The refusal of the MDF to withdraw some of its candidates in favour of Fidesz signalled imminent defeat. I can personally attest that at the critical moment Viktor Orbán was ready to accept an MDF candidate for the post of Prime Minister. In between the two election rounds I transmitted this message to Ibolya Dávid but she never responded. The opportunity to overturn the Socialists was missed, and Gyurcsány was re-elected Prime Minister.

When it was revealed in autumn 2006 that Gyurcsány, in a closed discussion with his parliamentary group, had admitted to lying about the economic situation of his country and that he had given false promises, all hell broke loose. Orbán demanded his immediate resignation. To empower this demand large demonstrations were held on the occasion of the fiftieth anniversary of the Hungarian uprising, which were suppressed by force by the police. Together with Viktor Orbán I addressed a demonstration of tens of thousands of people. We were both criticised by the Left for failing to uphold the ideals of the Hungarian Uprising. At the same time, the Party of European Socialists offered their full support to the lying Prime Minister. Since then Viktor Orbán has won a referendum against the government by an overwhelming margin, and Gyurcsány is in steep decline in the polls.

The Velvet Revolution

After the foundation of the Czechoslovak Republic in 1918 the Czechoslovak People's Party (CSL) was created; one of the very first Christian people's parties. It took part in nearly all governments, defended democracy and had a non-confessional name. During the conference at Munich (*Münchner Abkommen*), Czechoslovakia was abandoned by the British and French Prime Ministers, Neville Chamberlain and Édouard Daladier, and, with the help of Mussolini, Hitler annexed the Sudetenland to Germany. The CSL leader and Prime Minister of Czechoslovakia, Jan Sramek, opposed fascism and was therefore exiled from his country in August 1939. After the war the CSL won 16% of the votes and returned to government. However, following the putsch of Prague in February 1948, the country was taken over by the communists. Of the CSL's 46 Members of Parliament, 26 either emigrated or refused to collaborate and landed in jail or concentration camps. After the Velvet Revolution of November 1989 the collaborators deserted the CSL and a new, independent Christian people's party emerged. Together with a new Christian Democratic Party it founded the Christian and Democratic Union-Czechoslovak People's Party (KDU-CSL). On 7 June 1990 the party became a full member of the EUCD together with the Christian Democratic Movement of Slovakia (KDH). It participated in the centre-right government that ruled until 1998. Thereafter it remained in opposition for four years but since 2002 it has again borne the responsibility of government.

The unsung hero who founded this party was Josef Lux. On 5 April 1992 he invited me to Prostejov-Brno where I gave a speech at their congress on the topic, "only personalism can set humans free". Unfortunately, Lux died young in 1999. The leadership of the party was inherited in turn by Cyril Svoboda, Miroslav Kalousek and Jiří Čunek.

It is no wonder that the KDU-CSL remains a strong European-oriented party, even in the present centre-right government led by the partly Eurosceptic Civic Democratic Party (ODS) founded by President Václav Klaus and currently run by Prime Minister Mirek Topolanek. Klaus became notorious for saying that "there are not enough borders in Europe", which is in complete contradiction to the Treaty of Rome that envisaged four fundamental freedoms: free of movement of goods, persons, services and capital. With regard to Europe, Klaus is the complete opposite of his predecessor, Václav Havel. Havel became the symbol of resistance to Communism after the suppression of Alexander Dubček's "Prague Spring" in 1968. Because of his brave opposition Havel became the key figure of the Velvet Revolution. As the "philosopher president" he continued to inspire his people from 1990 to 2003, personifying his political policy in a different manner and with a deep vision of Europe.

During the summer of 2008 Havel stated in an interview for *l'Express* that European integration must continue and that it is irreversible. He reproached Klaus for his provocative talk after the May 2008 Irish referendum, when the latter declared that "the Treaty of Lisbon is dead". According to Havel, "in ten to fifteen years we will be in need of a European Constitution, a short text understandable by everyone, a text that children can easily learn at school. The rest is food for academics. Europe suffers from a lack of leadership and today the European Union resembles a bureaucratic institution that is occupied merely with technical and administrative matters. The "leader" cannot merely be the guardian of customs, quotas, tariffs etc. The Union has a cultural, historical, traditional and spiritual dimension and therefore the main task of the "leader" should be to focus on and emphasise this core" (*L'Express*, 14 August 2008).

With regard to the ODS, Mirek Topolanek, who is a pragmatic politician, tried to move out of the shadow of Klaus prior to his election as Prime Minister. Part of this attempt was to bring his party closer to the EPP and to reinforce the pro-European wing of the ODS. The ODS is loosely allied with the EPP as part of the "ED" sub-group (together with the British Conservatives) in the EPP-ED Group in the European Parliament. Topolanek, who has participated many times in the EPP Summit, asked to meet with me privately prior to the 16 June 2005 EPP Summit to discuss EPP membership options for the ODS. The discussion was very positive and I sensed that his effort was genuine. In fact, a few months later he invited me to address the sixteenth anniversary congress of the ODS in Brno on 26 November 2005, which I gladly accepted and where I was warmly received. But it was obvious that Klaus and his supporters were not very amused with this rapprochement and did everything in their power to sabotage the process. Sadly, Topolanek has given in to this pressure, especially after his election as Prime Minister where he now leads a very fragile government. Topolanek even went as far as to co-sponsor with David Cameron the so-called Movement for European Reform, to the delight and appeasement of the Eurosceptic wing of the ODS. Nevertheless, I still have some hope for future relations between the EPP and the ODS since, as polling has repeatedly shown, the overwhelming number of ODS voters support the European project.

... and Velvet Separation

Reflecting on the situation in my own country, I gave a speech in Bratislava on 23 May 1992 entitled "Belgian and Czechoslovak Federal Experiences: Union through Federalism or Cooperation?" Was this an expression of true belief or naivety? In June 1992 Vladimir Mečiar won the elections in Slovakia with a left-nationalist programme, while in the Czech Republic the liberals and the conservatives of Václav Klaus, who wanted to move forward with economic reforms, obtained the majority of votes. Their differences were too great to form a coalition. Klaus and Mečiar, the Czech and Slovak Prime Ministers, met in Brno in Villa Tugendhat, a modern masterpiece by German architect Ludwig Mies van der Rohe, and concluded that the federation was a thing of the past.

The separation was worked out in secret deliberations between the Parliaments. There was no referendum. For independent Slovakia, the government of the authoritarian nationalist Mečiar was pernicious. He refused reforms, maintained bad relations with NATO and the European Union and drove his country into international isolation because of his gross manipulation of the election legislation and the infringement and violation of minority rights. Five years later, in October 1998, Mikuláš Dzurinda put Slovakia back on track. With his Slovak Democratic Coalition (SDK) he obtained 42 seats and was able to form a stable government. In January 2000 he formed a new party, the Slovak Democratic and Christian Union (SDKU). In 2002 he won a second mandate and formed a new government with the two other EPP member parties, the Christian Democratic Movement (KDH) and the Party of Hungarian Coalition (SMK) and a junior partner, New Citizen (ANO).

During these past eight years Mikuláš Dzurinda has changed the image of Slovakia entirely through political and economic reforms and membership in the European Union and in NATO and has secured harmonious cooperation with minorities. Together with the Christian Democrats of the KDH and the Hungarian Coalition Party of SMK, he has transformed his country into a big success story, a role model for Central and Eastern Europe. After the elections of 2006 these three parties could again have formed the government. However, because of differences between the three, governance was passed over to the socialist/populist Robert Fico, who formed a coalition together with the autocrat Mečiar and Ján Slota of the Slovak National Party, well known for its xenophobic views and hostility towards ethnic minorities. Again it was proven that only understanding and cooperation between the members of the same political family can lead to a lasting result.

The Beginning of the End of Communism

When I first visited Poland in June 1970 the country was ruled by Edward Gierek. Under his leadership the communist regime became temporarily more liberal, expanding some personal freedoms, but most of the time persecution of the democratic opposition persisted. In 1978 Lech Wałęsa began with other activists to organise free, non-communist trade unions. In 1980 he led the strike at the Gdansk shipyard, which gave rise to a wave of strikes all over the country. The primary demands were for workers' rights. The authorities were forced to capitulate and to negotiate the Gdansk Agreement of 31 August 1980, which gave the workers the right to strike and to organise their own independent union, Solidarity.

The Catholic Church supported the movement, and in January 1981 Wałęsa was received by Pope John Paul II in the Vatican. Wałęsa himself has always regarded his Catholicism as a source of strength and inspiration. The country's brief enjoyment of relative freedom ended in December 1981, when General Wojciech Jaruzelski imposed martial law, "suspended" Solidarity, arrested many of its leaders and confined Wałęsa to a country house in a remote location.

In November 1982 Wałęsa was released and, although kept under surveillance, he managed to maintain lively contact with Solidarity leaders in the underground.

Martial law was lifted in July 1983, and in October 1983 the announcement of Wałęsa's Nobel Prize raised the spirits of the underground movement. Despite an attempt by the government to crack down on the anti-communist sentiments, the opposition had gained too much momentum and it became impossible to hold off change any longer. In addition, there was fear of a social explosion due to the economic malaise and runaway inflation that had eroded standards of living and deepened public anger and frustration. In September 1988 a secret meeting was held which included, among others, the opposition leader Lech Wałęsa and Interior Minister Czesław Kiszczak. They agreed on holding the so-called Round Table talks to plan the course of action to be undertaken in the country. The talks began on 6 February 1989 and were co-chaired by Wałęsa and Kiszczak. They included the Solidarity opposition faction and the coalition government faction.

During Easter 1989 I had a meeting with the last communist Prime Minister of Poland, Mieczyslaw Rakowski. During that same visit President Jaruzelski tried to convince me that martial law and the elimination of Solidarity in December 1981 had been necessary to prevent Soviet intervention. Years later, Rakowski published his memoirs under the title, "It Started in Poland: The Beginning of the End of the East Block" (this was published in German as *Es begann in Polen. Der Anfang vom Ende des Ostblocks*, Hoffman und Campe, 1995). Rakowski emphasises that the Polish opposition functioned as the motor for the end of the communist regime and that he was in favour of reform, an attitude shared by a part of the establishment.

The Communists hoped to co-opt prominent opposition leaders into the ruling group without making major changes in the political power structure. In reality, the Round Table talks radically altered the shape of Polish society. The result was the holding of parliamentary elections which led to the formation of a non-communist government under the leadership of Tadeusz Mazowiecki, one of the leaders of Solidarity and the first non-communist Prime Minister in Central Europe after the Second World War. The events in Poland precipitated and gave momentum to the fall of the entire Communist bloc: the Yalta arrangement soon collapsed.

Lech Wałęsa became President of Poland on 9 December 1990. I received him in Brussels on 1 July 1991 and already then he had opened the prospect of his country joining the European Union. However, at the end of 1995 he lost his second bid for re-election. In 1992 I became acquainted with the identical twin brothers Lech and Jarosław Kaczyński. They were our main contacts when the EUCD held its twenty-fourth congress in Warsaw on 21 June 1992. In my speech "The European Community, Preview of the Europe of Tomorrow" ("La Communauté européenne, préfiguration de l'Europe de demain", published in *L'une and l'autre Europe*, Éditions Racine, 1994, pp. 131–133) I mentioned that "In the new global context characterised by multipolarity, a new Europe has to be constructed with new structures capable of facing the challenges of tomorrow. It is clear that the problems of the new Europe can only be resolved in their entirety".

In 1997 Jerzy Buzek was elected Prime Minister, first of the centre-right AWS-UW coalition (Akcja Wyborcza Solidarność [Election Action Solidarity]–Unia Wolności [Union for Freedom]) until 2001, and then of the AWS minority government. Reforms were undertaken in the areas of education, pensions and regional administration that were necessary but because of which the party lost considerably in popularity. In the elections of 23 September 2001, Buzek lost to the SLD, the former communists, who garnered 41% of the votes. Prime Minister Jerzy Buzek obtained less than 5% of the votes and fell under the minimum threshold for re-election. But in June 2004 Buzek got his revenge and was elected Member of the European Parliament, running for election only on the popularity of his name and his direct contact with the voters. He received the record number of votes in the whole of Poland: 173,389 (22.14% of the total votes in this region).

In 2001 the Kaczyński brothers founded the conservative party Prawo i Sprawiedliwość (Law and Justice), or PiS. At first they were in favour of membership in the EPP, and their MEPs became member of the EPP-ED Group in the European Parliament. However, Jarosław Kaczyński used a conflict with the German "Bund der Vertriebenen" to cancel the cooperation of his party and to withdraw their MEPs from our group. Since then, the two brothers have been on a Eurosceptic course and acted strongly on this tendency when they came into government. In December 2005 Lech Kaczyński was elected President of Poland and soon after his brother Jarosław Kaczyński became Prime Minister. Two years later, in the 21 October 2007 elections, PiS was beaten by EPP member party Platforma Obywatelska (PO, Civic Platform) with the help of the Polskie Stronnictwo Ludowe (PSL, Peasants' Party), also an EPP member.

In November 2007 PO leader Donald Tusk became Prime Minister and formed a strong government with PSL leader Waldemar Pawlak. Donald Tusk was one of the founders of the Civic Platform and from June 2003 he took over as party leader. In the 2007 elections, he received more than 534,000 votes, which is the best individual result in the electoral history of the Third Polish Republic, and his Civic Platform won the elections with 41% of the vote. Donald Tusk represents a new Poland that is focused on the important values of Western civilisation. On the strength of these ideals he was able to convince the younger generations of his country in a remarkable election campaign. He is our great hope, therefore, for a real European Poland.

Baltic Strength

The reunification of Europe would not have been complete without the independence of Estonia, Latvia and Lithuania from the former Soviet Union. The three small Baltic states were quick to implement the rule of law, to develop vibrant free-market economies and to firmly establish parliamentary democracies. These rapid reforms are largely credited to the centre-right political parties that came to power at critical moments in the transformation of these countries.

One characteristic example is the government of Mart Laar in Estonia. At a rather young age (he was born in 1960), Laar and his Pro Patria Union came into government in two important periods – 1992–1994 and 1999–2002 – and as Prime Minister he managed to lead Estonia through the lightning economic reforms that won praise and ultimately laid the groundwork for rapid economic growth and acceptance into the European Union. Laar's economic reforms led to the Baltic Tiger period that started for Estonia after 2000. The reform process continued under Juhan Parts and his Res Publica party when Parts became Prime Minister in 2003–2005. It is important to note that both Pro Partia Union and Res Publica joined the EPP, and both Laar and Parts have had a very active presence in the EPP Summits. After the two parties merged in 2006, Mart Laar came out of political retirement and took over the leadership of the new party. Today Laar remains one of the most respected and active political figures in the EPP.

In Latvia, both of our EPP member parties have also been in government and left their mark. Jaunais Laiks (New Era), led by Einars Repše, became part of the government in 2002 and Repše became known for his outspoken fight against corruption and tax evasion. In December 2004 a new government led by Tautas partija (People's Party) leader Aigars Kalvītis pushed on with the reforms and set the pace for Latvia's role as a new EU Member State. In his three years as Prime Minister, Kalvītis strengthened the profile of Latvia as a solid and reliable EU partner – it is not surprising that Kalvītis never missed a single EPP Summit during his time as Prime Minister. Also, Kalvītis' support for Andris Piebalgs as Latvia's member in the Barroso Commission should not be underestimated. Piebalgs surprised many with his impressive work as Commissioner for Energy. It is safe to say that he is one of the best Commissioners the EPP has had.

In Lithuania, the person who left his mark on behalf of the country's independence and helped trigger the independence of many former Soviet republics is Vytautas Landsbergis. As leader of the pro-independence movement, he won the 1990 elections and soon after declared the independence of Lithuania. Landsbergis became the President of Lithuania and managed to resist the economic blockade of the Soviet Union. His brave actions inspired many others working for freedom in the former Soviet Union to follow the same course. In 1993 Landsbergis founded the centre-right party Homeland Union, which won its first term in office in the 1996 elections. His successor in the Homeland Union, Andrius Kubilius, came to prominence in 1999 when he became Prime Minister but, unfortunately, lasted only one year in government. Yet in 2008 he made an impressive comeback when he won the October 2008 parliamentary elections and became Prime Minister for a second time. I expect that he will stay in office for a full term, since he managed to forge a strong centre-right coalition. Prior to his election victory, Kubilius successfully accomplished the merger of the two Lithuanian EPP member parties – the Homeland Union and the Lithuanian Christian Democrats – and the new party is now the largest in the Seimas, the Lithuanian Parliament. Kubilius, both as opposition leader and now as Prime Minister, has been a loyal participant in EPP events.

The Mediterranean Island States

The third-largest island of the Mediterranean Sea, Cyprus is a country that has always been considered part of Western Europe. For decades the Republic of Cyprus had been developing a stable democratic system and a vibrant free market economy but, ironically, the Eurosceptic radical Left party AKEL had always managed to control about a third of the Greek-Cypriot electorate. AKEL resisted Cyprus joining the EU and it is not surprising, therefore, that the then AKEL-backed President of Cyprus, George Vassiliou, decided to apply for EU membership only in 1990, after the fall of the Berlin Wall. In 1993, Glafkos Clerides, founder of the centre-right Democratic Rally (DISY) was elected President of Cyprus.

Clerides has always been a convinced European and this is also why his party, DISY, has been and continues to be the most pro-European party of Cyprus. Clerides also understood that a European future for Cyprus would help the reunification process of the island, since accession would offer an important incentive for Turkish Cypriots to accept a compromise solution and would also put pressure on Turkey to end its military occupation of the north. When Clerides was elected President of Cyprus, he made the goal of EU accession a top priority and marshalled all of his resources to achieve this aim. In order to do so, Clerides quickly understood the importance of joining the EPP, and one of the first things he did was to ask his successor at DISY, Yiannakis Matsis, to apply for EPP membership. In 1994, DISY became an associate member of the EPP, and during the ten-year presidency of Clerides he was a regular participant at the EPP Summit. In these ten years, he managed to successfully complete the accession negotiations; this was largely due to the political acumen and successful governance of Clerides.

Unfortunately, the same cannot be said for the current Cypriot President and Secretary-General of AKEL, Dimitris Christofias, whose Euroscepticism threatened to derail the Lisbon Treaty even further when his party voted against the Treaty in July 2007. If it had not been for the support of the DISY and its current leader Nikos Anastasides, the Lisbon Treaty would also have failed in Cyprus and would have put Europe in an even deeper crisis. In fact, Anastasiades has worked very vigorously in the EPP since his election as DISY President in 1997 and also took the initiative to host on behalf of the EPP the "Middle East Observatory".

Another leader for whom I have high regard is Eddie Fenech Adami, the current President of Malta and the country's longest-serving Prime Minister since independence. Fenech Adami is staunchly pro-European and is almost single-handedly responsible for Malta's accession to the EU. In 1996, when his Nationalist Party (PN) lost the elections, Malta's European Union application was put on hold by the Eurosceptic Labour government. Fenech Adami's strong opposition managed to bring down the Labour government in 1998; as Prime Minister, he immediately re-activated the EU accession negotiations. Fenech Adami's crowning achievement came when he was able to win the March 2004 referendum on Malta's EU accession with 53.65% of the votes. Once again, the Labour Party campaigned vigorously against EU accession and failed. Actually, I have never really understood why the Maltese Labour Party is a member of the Party of European Socialists – the EPP would never accept as a member a party that campaigns against Europe. The PN, on the other hand, has proven time and time again its pro-European commitment and has also been actively involved in the EPP. When Fenech Adami was Prime Minister, he always attended the EPP Summits. The same holds for his successor, Prime Minister Lawrence Gonzi, who is very actively involved in our discussions. I am also impressed with Deputy Prime Minister and Foreign Minister Tonio Borg, who is engaged in our work and in our EPP Foreign Ministers' meeting in particular.

From a Social Democratic to a People's Party

The most authentic Romanian that I met after the December Revolution of 1989 was Corneliu Coposu. Born in Transylvania, a part of Austria-Hungary at the time, he joined the Romanian National Party (PNR), a group dominated by Greek-Catholic politicians, and engaged in local politics with the PNR's direct successor, the National Peasants' Party (PNȚ). In 1940 he became political secretary to Iuliu Maniu, the leader of the PNȚ, who was Prime Minister and a decisive factor in Transylvania's union with Romania. In this capacity Coposu accompanied Maniu to all treaty negotiations between Western negotiators and the leaders of political parties in the anti-Hitler coalition during the war.

After Iuliu Maniu's arrest on 14 July 1947, Coposu remained in prison for nine years without being tried, until 1956 when a trial for high treason was staged. He was convicted according to the communist practice to hard labour for life for "high treason against the workers' class and crimes against social reforms", and underwent eight hard years of solitary confinement. His wife, Arlette Marcovici, was arrested in 1950 and falsely charged with and tried for espionage; after spending fourteen years in prison she died in 1965. As a consequence of the brutal treatment applied to political prisoners, Coposu became seriously ill, and in April 1964 he was freed after seventeen years in prison. He went to work as an unskilled worker on construction sites, permanently followed and periodically brought in for questioning by the *Securitatea*.

In 1987 Corneliu Coposu joined the underground activity of the Peasants' National Party. As President he announced in December 1989, through a published manifesto, the re-entry of the party into public life. Later on he transformed it into a Christian Democratic party, the PNȚCD. Understanding that the revolution of December 1989 had been stolen by the former members of the old regime, he would take on the hard task of leading the opposition in post-communist Romania, becoming the target of a virulent press campaign. The idea to form the Romanian Democrat Convention (*Convenția Democrată Română*, CDR) was Coposu's; he had proposed it since 1990. He was the chair of this Convention from 1991 to 1993. He realised his ambition to present a candidate capable of defeating the post-communist President Ion Iliescu. His candidate was Emil Constantinescu and I witnessed how Coposu lobbied for support for Constantinescu with the Christian Democratic parties in Western Europe. Coposu died on 11 November 1995 and is buried in the Catholic cemetery of Belu. He was not given a state funeral, but hundreds of thousands of Romanians came to accompany him on his last journey.

Constantinescu was elected President of Romania in November 1996 and appointed Victor Ciorbea, the mayor of Bucharest, as Prime Minister. The year 1997 began with great expectations. Initially support for the new government was high, even as it initiated its shock therapy. However, the reforms proved difficult: given the slow pace of privatisation and the stagnation during the previous government, the attempt to restructure state industries was difficult. The delayed reforms can be explained by the lack of homogeneity and consensus among the members of the coalition, which was formed of three political groupings: the centre-right CDR, the socialist-leaning Uniunea Social-Democrată (USD) and the Democratic Union of Hungarians in Romania (RMDSZ). Widespread disagreement and tension surfaced within each of the three groups, as well as among them, and nearly every political formation was plagued by infighting and rifts.

This crisis revealed that a heterogeneous coalition, broadly in agreement on aligning the country with the West but divided over personal rivalries and policy details, lacked control over key parts of a bureaucracy unreformed since communist times. Viktor Ciorbea resigned in March 1998, both as Prime Minister and as mayor of Bucharest. The results of the governments under the leadership of his CDR-successors were disastrous. Romanians were strongly disillusioned with the major parties and politicians. Many viewed them as a separate caste whose primary aim was to protect special corporate interests rather than the common good. A disenchanted Emil Constantinescu, who lost popularity and had failed to fulfil his reforms, announced that he would not run for a second term and withdrew from political life in November 2000. Only the government's foreign policy can be seen as a strong point. It adopted a pro-Western stance, and early in its mandate launched a diplomatic offensive to improve the image of Romania abroad. Joining NATO and the European Union were proclaimed Romania's top foreign policy priorities.

After his time as Prime Minister, Victor Ciorbea attempted for months to revive the PNT‚CD. I supported him to the utmost with friendship and loyalty. But when it became clear from all the polls and disastrous election results that the party could not be resurrected, I was forced to explain to him using Biblical language that "it was time for a John the Baptist". In 2004 he was succeeded by Gheorghe Ciuhandu, the Mayor of Timis‚oara, who was one of the losing candidates in the 2004 presidential elections and who was replaced in turn by Marian Petre Milut‚ in 2007. Sadly, today the PNT‚CD is only a marginal party in Romania.

Current Romanian President Traian Ba˘sescu was a member of the pre-1989 Communist Party (PCR), and after its downfall he claimed that he had joined the PCR in order to promote his career as a Merchant Navy officer. After the revolution he became a member of the National Salvation Front (FSN). In 1992 the FSN split into two factions – the Social Democratic Party of Romania (PDSR, later PSD), led by Ion Iliescu, and the Democratic Party (PD), led by Petre Roman. Ba˘sescu joined the PD and succeeded Roman as its President in 2001. After taking over the PD, he gradually began shifting the party in a centre-right direction. In 2004 Ba˘sescu won the presidential election. During a TV debate with Prime Minister and PSD President Adrian Năstace, he caught his main opponent off guard with a cynical remark: "You know what Romania's greatest curse is right now? It's that Romanians have to choose between two former Communist Party members". Running on a strong reform and anti-corruption platform, Ba˘sescu's victory was characterised as Romania's Orange Revolution, referring to the orange colour that was heavily used in the campaign. He honoured an agreement he had made with the National Liberal Party (PNL) and appointed its leader, Ca˘lin Popescu-Ta˘riceanu, as Prime Minister, but their relations gradually soured.

In March 2005 President Ba˘sescu informed me that the Democratic Party aspired to join the EPP. On 25 June the party's leader, Emil Boc, mayor of Cluj-Napoca, the largest city in Transylvania, called an extraordinary national congress to realise this transformation from a Social Democratic to a People's party. EPP membership was granted by the EPP Political Bureau on 19–20 September 2005. President Ba˘sescu attended the EPP Summit for the first time on 23 March 2006. This transformation was, politically, a striking moment: it was the first time that a political party with post-communist origins had adopted the basic principles and programme of European Christian Democracy. In January 2008 the Democratic Party merged with the Liberal Democratic Party and created an even broader centre-right party, the Democratic Liberal Party (in Romanian, Partidul Democrat-Liberal, PD-L). Although as President of Romania Ba˘sescu cannot formally hold any party function, people closely associate the PD-L with him. The PD-L leadership has already stated publicly that it will support Ba˘sescu's candidacy in the next presidential elections. And in the November 2008 parliamentary elections, the PD-L won the most seats in both the Chamber and the Senate, but not enough to form a government on its own. Thus, the PD-L was forced to form a grand coalition with the Social Democratic Party, and PD-L chair Emil Boc became the Prime Minister.

Hope for Change and the Rule of Law

A similar evolution took place in Bulgaria. Ivan Kostov entered politics after the collapse of the Berlin Wall and the fall of the communist dictator, Todor Zhivkov. In 1990 the centre-right Union of Democratic Forces (UDF) was created as the major anti-communist party. Kostov was elected chair in 1994 and, in a crushing victory in the May 1997 elections, the UDF got 55% of the vote. Kostov became Prime Minister, his cabinet being the first post-communist government to serve its full four-year term. Long-delayed economic reforms were carried out with the privatisation of state-owned enterprises on a large scale. The country started accession talks with the EU, which Bulgaria eventually joined in January 2007.

Kostov and his Minister of Foreign Affairs, Nadezhda Mikhailova, proved to be a successful pair in carrying out these reforms. However, the elections of June 2001 were sobering. The reformist policies were widely criticised, notwithstanding the EPP's manifest and public solidarity. I succeeded in convincing our EPP heads of state and government to participate in an EPP Summit in Sofia on 5 April 2001. The event was organised with extra pomp and ceremony and was broadcast live on television. At the reception following the event, former King Simeon, of Saxe-Coburg and Gotha (or Simeon II of Bulgaria), also turned up. Some naive politicians were convinced that he would support Kostov in the election campaign.

The next day Simeon announced the formation of a new political party, the National Movement Simeon II (NMSII), dedicated to "reforms and political integrity". Simeon proclaimed that in 800 days the people would feel positive effects and would enjoy significantly higher standards of living. The NMSII won a large victory in June 2001, and Simeon formed a coalition with the Turkish ethnic minority party Movement for Rights and Freedoms (MRF). During his time in power, Bulgaria joined NATO, but the country remained mired in poverty, deteriorating public services, government corruption and organised crime. In the 2005 elections, Simeon's party came in second and participated in a grand coalition government with the Bulgarian Socialist Party and the MRF.

The UDF and the entire centre-right opposition grew ever more hopelessly divided. Kostov broke away from Nadezhda Mikhailova who, as EPP Vice-President at the time, could have provided new impetus to the UDF. Kostov left the UDF and formed the Democrats for a Strong Bulgaria (DSB). The Bulgarian centre-right became depressingly fragmented.

Thankfully, the salvation of the centre-right came about through a new political movement, citizens for the European Development of Bulgaria (GERB). Although GERB is headed by the former deputy mayor of Sofia, Tsvetan Tsvetanov, the party's actual leader is the charismatic mayor of Sofia, Boyko Borisov. GERB won the 2007 European elections with 21.69% of the vote, and its five MEPs joined the EPP-ED Group. A few months later GERB became a member party of the EPP. Borisov offers hope for change and for strengthening the rule of law; he is the only possible alternative in the fight against corruption and organised crime. I expect that GERB will win both the European and national elections of June 2009.

To Be Right and to Be Proven Right

There is no other country in the Western Balkans that comes as close to the European Union's objectives as Croatia. The country definitely deserves to become the twenty-eighth Member State of the Union. But it was difficult at first, as some governments lacked political courage in the long discussions on the start of the negotiations. Thanks to my excellent relationship with Prime Minister and HDZ leader Ivo Sanader and my work in Bosnia-Herzegovina, I was convinced that Croatia was fully cooperating with the International Criminal Tribunal for the former Yugoslavia (ICTY). On 15 March 2005, I chaired in Brussels a meeting of EPP Foreign Ministers in order to avoid a deadlock in beginning Croatia's accession negotiations. We did not succeed.

Opposition to the start of the negotiations was fed by the absurd declarations of Carla Del Ponte, the Swiss head prosecutor of the International Criminal Tribunal for the former Yugoslavia (ICTY), who stated that the Vatican was protecting the Croatian war criminal Ante Gotovina. This former Lieutenant General of the Croatian Army was indicted in 2001 for a "joint criminal enterprise", the attempt to expel Krajina Serbs from Croatia in 1995. The United Kingdom, the Netherlands and other Member States made his surrender a precondition for Croatia's accession. This stance was criticised by the Croatian government, which claimed that it did not know of Gotovina's whereabouts, that he was probably outside the country and that it was doing all it could to bring him to justice. Accession negotiations with the EU, scheduled to start on 17 March, were postponed.

But I did not give up. I undertook various actions that could lead to a breakthrough. One was the 26 August 2005 letter I wrote to Tony Blair, who was then President of the European Council, in the name of nine Prime Ministers, eight of whom were from the EPP: Silvio Berlusconi, Mikuláš Dzurinda, Lawrence Gonzi, Janez Janša, Jean-Claude Juncker, Aigars Kalvītis, Kostas Karamanlis, Wolfgang Schüssel (plus Andrus Ansip from Estonia, even though he belongs to the European Liberals):

> Mr. President
>
> This autumn, important decisions will have to be taken by the European Union. Recognising the importance of regaining and reviving confidence, values, mission and popular acceptance, the Union will have to continue to address the main issues that have dominated the internal as well as the public debates in recent months. It is a period of self-reflection for the European Union.
>
> At the same time, the global role of the EU must be maintained and strengthened. We appreciate the determination, focus and earnestness of the EU Presidency in attending to these pressing issues. In particular, we fully support your efforts to maintain the momentum of the European debate in all areas.
>
> In this context, we firmly believe the negotiations with Croatia must be opened. These negotiations will provide the best support for the ongoing reforms and the strengthening of democracy, stability and prosperity in Croatia and the entire Western Balkan region, in line with the Thessaloniki agenda.
>
> Croatia deserves our full support. The responsibility and credibility of Croatia's government and its forward-looking policy in the region must be acknowledged. The time has come for accession negotiations with Croatia to be opened. We are convinced that the fulfilment of the Croatian government's Action Plan related to the EU Council of Ministers decision of 17 March 2005 now opens the way to start accession negotiations with Croatia this September.

At first I merely received an acknowledgement of receipt from a staffer in Blair's cabinet. After the Council finally had decided in October 2005 to start negotiations, Blair replied enthusiastically and thanked me for my letter of 26 August about the global role of the EU under his presidency and, more specifically, Croatia's accession negotiations with the EU.

The EPP and Ivo Sanader were finally proven right when Gotovina was arrested by the Spanish police in Tenerife on 7 December 2005. Soon after, Ms Del Ponte resigned from ICTY and disappeared from the public eye. Apart from pushing forward Croatia's EU membership process, Ivo Sanader has emerged as the leading political figure of his country. When he took over the leadership of the Croatian Democratic Union (HDZ), Sanader managed to transform it from a nationalist party dominated by the questionable legacy of Franjo Tudjman to a modern, pro-European centre-right party that wants to develop good neighbourly relations with all of its former rivals. As a result, Sanader led the HDZ to victory in the 2003 and 2007 parliamentary elections and is poised to make history as the Prime Minister of the twenty-eighth EU Member State. Nevertheless, the first years of Croatia's accession negotiations moved at a very slow pace. The Commission was clearly not in a hurry: after more than two years, two chapters were opened and temporarily closed again, while fourteen more were opened for the first time. During a dinner in my office on 20 February 2008 Ivo Sanader and I attempted to convince the President of the Commission to make Croatia's accession negotiations a higher priority.

Sanader himself took a major political risk: Croatia's parliament approved his proposal to give EU Member States the right to fish in a sensitive zone in the Adriatic Sea. Previously, Parliament had proclaimed the area a protected fishing and ecological zone – a so-called ZERP – saying it aimed to limit fishing there in order to protect marine life. This decision particularly angered Croatia's neighbours Slovenia and Italy. "Both issues are important for Croatia: they both are of national interest", Ivo Sanader told Parliament, "but EU accession is an 'absolute national interest'." The decision was passed on 13 March 2008 at one O'clock in the morning by a small majority: 77 parliamentarians voted in favour, with nine against, while the rest of the lawmakers in the 151-seat body preferred to abstain. During the European Council of that same day the Commission welcomed the decision and declared that it intended to conclude the accession negotiations with Croatia before its own mandate expired in October 2009.

This important political step was also of great importance for the relations between Croatia and Slovenia. The two ex-Yugoslav countries, although led for many years by EPP Prime Ministers, often had tense relations due to border disputes. On 15 February 1991 I met Alojz Peterle for the first time. Since 1990 he had been Prime Minister of Slovenia, and to the great astonishment of my own diplomats, I received him in my residence for an official audience before Slovenia declared independence on 25 June 1991. Peterle was also the leader of the New Slovenia–Christian Democratic Party (Nova Slovenija–Krščanska Ljudska Stranka). After many important ministerial positions he was elected Member of the European Parliament and also Vice-President of the EPP. New Slovenia's heritage lies in the rich tradition of the Christian Socialist movement that developed in the nineteenth century (1892) and in the historic (prewar) Slovenian People's Party.

Another key factor in Slovenia's political development was the foundation of the centre-right Slovenian Democratic Party (SDS) in 1989 and the role that Janez Janša would play. Under his leadership, in the elections of October 2004 the SDS became the largest political party in Slovenia. Janez Janša was elected Prime Minister and formed a coalition government of EPP member parties: the SDS, NS-KLS and Slovenian People's Party (SLS).

This government proved to be very efficient and played a remarkable role in the first half of 2008 during its Presidency of the European Union, confronted by the Kosovo issue. Janša would surely have won the September 2008 elections had it not been for the hateful and unfounded campaign against him and his government. Nevertheless, the SDS remained the strongest party, minus only one seat. For New Slovenia the result was very disappointing, since it would no longer be represented in Parliament. The party is now left with two MEPs, Ljudmila Novak and Alojz Peterle, on whose shoulders falls the heavy task of reinvigorating the NS-KLS.

The Call of Bosnia

A large part of my activities was focused on countries with the long-term goal of joining the European Union. I witnessed at first hand the complexities in these countries when I chaired a committee preparing the reform of the police in Bosnia and Herzegovina. Since the Dayton Agreement the European Commission has been closely involved in the stabilisation of the country and has attempted reforms in crucial areas such as the reorganisation of the national police force. Chris Patten, Commissioner for External Affairs, offered me the mandate to chair the committee on the country's police reform, in view of my Belgian institutional past and my knowledge of the Bosnian political spectrum. I accepted, and it was a unique experience to apply on the ground the principles that we defend from Brussels.

The police in Bosnia and Herzegovina proved to be affected by corruption and was afflicted by political interference. There were far too many policemen and the cooperation between the forces of the different political entities was non-existent. This was proven beyond a shadow of a doubt in the report prepared by Bernhard Prestel, a prominent German expert whom I had already learned to appreciate when I was Prime Minister of Belgium. His report became the basis for the police reform. Apart from experts from Britain and Ireland, the committee was composed of local professionals involved with the police: the Bosnian Minister for Security, the Ministers for the Interior from Republika Srpska and from the Bosnian-Croat Federation, the Attorney-General, the Chief of National Security and so on. During a few fact-finding missions I tried to assess the situation on the ground with my own eyes and to learn to know the people and their problems.

In order to organise the police efficiently and effectively I proposed to create a double structure along the Belgian model: one national police and various local police zones provided with a large measure of autonomy in their daily work. Central to this structure was that the territorial partitioning be on the basis of efficiency and not on the basis of ethnicity. This was unacceptable to the Republika Srpska but, thanks to their abstention, consensus was reached on my proposals. It took until 2008 before the various parliaments in Bosnia and Herzegovina adopted the essential elements of my proposal.

My mandate in Bosnia and Herzegovina convinced me that the real future for the country lies in Europe. After the agreement on police reform the European Commission could finalise the Stability and Association Agreement, an important step before the country can receive the status of candidate Member State. Only the European Union can provide lasting stability in a country torn by "hyper-nationalism". This country is representative of the problems in the Balkans in the political, economical, social, cultural and religious fields. But stabilisation and the future of the whole of the Balkans region, the "time-bomb of Europe", lies in the Union. I see no other alternative. This region of European countries will have to adhere to the EU in the long run. It is a fact that the

European public has enlargement fatigue. Therefore there is a need for political leaders who have the courage to explain this to our citizens. Do we have the leadership and courage to bridge this gap? Are our leaders and governments aware of this huge challenge?

"Across the Country of the Illyrians"

This title, taken from Karl May, refers to Albania, the ancestral land of the Illyrians, the people that escaped the domination of the Romans, kept their own language and would later be part of the Ottoman Empire and therefore become mainly, but not exclusively, Muslim. Prior to the fall of Communism in Albania, the great writer Ismael Kadare sought asylum in France. During his exile, he stated that "dictatorship and authentic literature are incompatible ... the writer is the natural enemy of dictatorship".

On 17 December 1993 I took part in an EUCD mission to the country. Travelling there was difficult. The plane from Zürich landed at the small and badly maintained Tirana airport. We drove to the capital through fields and along numerous ugly heaps resembling termite mounds: small concrete artillery domes with which the communists had ruined the beautiful landscape of mountains in fear of attack. We stayed in Villa 31, a guesthouse in a green diplomats' district. I was given the enormous bedroom of former dictator Enver Hoxha. One wall was covered entirely with a giant picture – an autumn wood in flaming colours – of some twelve square metres. The Hoxha family clearly had taken delight in such exuberant bad taste because in the other rooms similar large scenes were also glued to the panelling.

Sali Berisha, a well-known doctor who specialised in cardiac surgery, got his first taste of democratic ideas while studying in Paris during the 1980s. He supported the student strike in December 1990, which turned political and obliged the government to allow the creation of new parties. The same month the Democratic Party of Albania (PD) was founded by students and intellectuals and in 1991, Berisha was elected its leader. After the first free elections, he became President in April 1992. During my conversation with Berisha in December 1993 he spoke enthusiastically of his government's economic shock therapy, with its radical privatisations, an end to inflation, the 200,000 cars in the country and its strong economic growth. Agriculture and the wine industry had been revived and telecommunications were being given priority. According to Berisha, "We had to catch up because we had become completely impoverished; this was apparent to all when our youth attempted to escape to Italy in such a pitiful way".

His government also allowed the setting up of Ponzi saving schemes. His party won the elections in May 1996, but the schemes failed, one by one, and in

December demonstrators took to the streets and accused the government of having stolen their money. Elections in June 1997 led to the victory of the Socialists. Berisha resigned from the presidency and moved into opposition. He lost again in 2001, but was able to cultivate a more moderate image for himself and his party. Widespread discontent with rampant corruption and the overweening arrogance of the Socialist Party resulted in his return to power in September 2005 after a marathon of legal challenges by the Socialists to the election results. In June 2007 he received George W. Bush, the first US President to visit Albania. During this visit, Bush repeated his staunch support for the independence of neighbouring Kosovo: "At some point in time, sooner rather than later, you have got to say 'enough is enough, Kosovo is independent'." (*New York Times*, 11 June 2007). Bush said any plan to extend talks on Kosovo, like those proposed by President Nicolas Sarkozy of France, must end with "certain independence".

The New Democratic Party (PDR) emerged in 2001 as a political formation from a reform movement within Berisha's Democratic Party. Genc Pollo was confirmed as its party leader. In the June elections that year it was established as the third force in the Parliament. The PDR was the first Albanian party to be admitted to the EPP as an observer member. Soon after, Berisha's PD also became an observer member. Happily, the two parties moved closer to each other, joined forces in the new government, and in October 2008 merged into one strong centre-right party in Albania. I hope that the renewal inspired by Sali Berisha and Genc Pollo will have a positive impact and will function as a role model for other parties in the Balkan region.

The European Choice of the People

After the fall of Slobodan Milošević as President of Yugoslavia in October 2000 we were looking for a democratic alternative. This was not easy: Zoran Djindjić, the leader of the Democratic Party, had studied in Germany and was friends with Willy Brandt. His party joined the Socialist International. We sought contact with Vojislav Koštunica, who responded with hesitation. He had been born into a legal family, and respect for the rule of law has remained a constant element in his career. He was both a Western liberal who believed in a free press, an independent judiciary and multi-party democracy, and at the same time a Serb nationalist often critical of the Western alliances but with no connection to the old Communist Party. He had lost his job as professor at Belgrade's Law School in 1974, after criticising Tito's communist government.

In 1989, Koštunica became one of the founders of the Democratic Party but left in July 1992 over opposing views on leadership and formed the Democratic Party of Serbia (DSS), of which he is still President. Supported by both nationalist and pro-Western voters, the democratic opposition backed him in the presidential election of September 2000. After the turbulent events of October, he was finally declared the winner and remained President of the Federal Republic of Yugoslavia until February 2003, when the state was replaced by Serbia and Montenegro and his position was abolished. In the meantime, the DSS joined the EPP, together with the G17 Plus.

Following the parliamentary elections in December 2003, in which the DSS emerged as the largest of the pro-European parties, Koštunica became Prime Minister in March 2004, albeit with the support of the Socialist Party of Serbia. However, as a result of the bad showing of the government candidate in the presidential elections, he announced that fresh elections should be expected by the end of the year. Since then, his minority government has maintained a frail coalition. In May 2007, after a brief crisis, he was sworn in for his second term as Prime Minister.

In February 2008, Koštunica called for new elections after the collapse of his coalition over European Union issues and Kosovo's declaration of independence. He stated that before its accession to the EU, Serbia and the EU must discuss Serbia's territorial integrity, and that Serbia could by no means sign the Stabilisation and Association Agreement (SAA) with the EU. These positions, taken by an EPP member party, were the cause of much tension. On 13 March the EPP Summit asked me to engage in a dialogue with the DSS. Together with Vice-President Viktor Orbán, I met with Mr Koštunica and his colleagues on 10 April 2008 in Belgrade. However, Boris Tadić, President of Serbia and leader of the Democratic Party, won the elections with a coalition in favour of the full integration of Serbia into the EU on the condition that Serbia would only join if its territorial integrity, including sovereignty over Kosovo, was respected. The G17 Plus also participated in this coalition and its leader Mladjan Dinkić became Deputy Prime Minister and Minister of Economy in the new government.

On 6 June 2008 I sent a letter to Mr Koštunica stating that after the elections of 11 May I had welcomed in a public statement the results for the DSS and that, despite the reservations he had expressed about EU foreign policy decisions during the emotionally charged campaign, I was convinced that he would accept the European choice that was clearly expressed in the elections by the citizens of Serbia. I added: "You are a great patriot and I know you will make the patriotic choice, which is the European choice. The leaders participating in the EPP Summit on 19 June expect a positive signal from you, indicating the willingness of yourself and your party to engage in the European choice of Serbia".

In Koštunica's answer of 17 June he confirmed that "the Democratic Party of Serbia does not have and has never had any issues concerning its European orientation nor with the European orientation of Serbia itself. The party program and actions of DSS, from its founding in 1992 to date, bear witness to that". He added concerning the controversy: "In essence it is, of course, the unlawful unilateral declaration of Kosovo's independence which is a brutal breech of the United Nations Charter and UN Security Council resolution 1244. I am sure that you will understand, Mr Martens, that this could not happen without having an effect on the internal as well as the international politics of the DSS, just as it would, under similar circumstances, no doubt have affected the policies of any other member of the EPP, from any other country". The discussion with the DSS was continued on 13 November 2008, when Mr Koštunica and a delegation participated in the EPP Presidency meeting. It remains our hope that before concluding our dialogue the DSS will unequivocally follow the European choice of the majority of the Serbian people.

In Search of an Identity

The dissolution of Yugoslavia also brought another independent state into being and with it more instability, both in the region and internally. The independence of the self-proclaimed "Republic of Macedonia" managed, merely by virtue of its name, to upset its southern neighbour, Greece. As is well known, Macedonia is also one of the largest Greek provinces, dating back to ancient Greece and to the Hellenistic culture established by Alexander the Great. This is why the EU and the UN have recognised this country under a provisional name, the Former Yugoslav Republic of Macedonia (FYROM), until it decides to accept a compromise solution.

More serious were the internal problems the country had to face after independence. Although the Slav-Macedonian population of the country is the majority, it is a multi-ethnic nation: more than a quarter of its two million inhabitants are Albanian and in addition there are Turks, Serbs, Roma and so on. As a result, an armed conflict between the country's security forces and the Albanian-led National Liberation Army erupted in the late nineties, since the Albanians demanded their rights be recognised on the same level as the rights of Slav-Macedonians. Even though at a certain point in the conflict there was a threat that the country could split, the two sides finally agreed on a compromise solution brokered by the EU and the US, and on 13 August 2001 the Ohrid Framework Agreement was signed.

As far as the involvement of the EPP is concerned, we established contacts at an early stage with the VMRO–DPMNE, the main centre-right party of the Slav-Macedonians, and to a lesser extent with the DPA of the Albanian community. These parties also participated in our Western Balkan Democracy Initiative led by then EPP Vice-President and ND leader Kostas Karamanlis. During that period, the country was led by VMRO–DPMNE leader and Prime Minister Ljubčo Georgievski, who was eager to establish relations with the EPP. It is interesting to note that in the framework of our early contacts, Georgievski also expressed willingness to compromise on the name issue and had a generally positive attitude towards Greece, but his main preoccupation at the time was to solve the armed conflict with the Albanians. In fact, it was Georgievski who finally signed the Ohrid peace deal in 2001; ironically, a year later he lost the elections to the Socialists.

Georgievski's successor in VMRO–DPMNE, Nikola Gruevski, continued to develop contacts with the EPP and his party finally became an observer member. Gruevski also managed to defeat the Socialists in the July 2006 elections, bringing his party back into government. During the membership review of VMRO–DPMNE, Gruevski gave written assurances that in the spirit of EPP values he was committed to developing good neighbourly relations. Nevertheless, unlike Georgievski, Gruevski opted to antagonise Greece by deciding, for example, to name Skopje's airport and the country's main highway "Alexander the Great", and by not accepting recent UN proposals for resolving the name dispute, namely, that FYROM adopt "Upper Macedonia" or "New Macedonia".

Many have criticised me for spending too much time with all these small countries and following all these complicated problems. Yet I continue to defend the importance of the EPP's being engaged in this fragile region. Let us not forget that, unlike the break-up of Czechoslovakia, the dissolution of the Socialist Federal Republic of Yugoslavia was very violent and thousands lost their lives. The region to this day continues to be unstable. I think it would be a mistake if the EPP were to turn its back on these fragile Western Balkan countries despite their numerous internal and regional problems. All like-minded parties of the region need to have a common political home, and we must offer them at least the hope that one day their country will join the European Union.

The Breadbasket of Europe

The search for a partner in Ukraine is in many ways comparable to what happened in Central and Eastern Europe. Already long before there was talk of the Orange Revolution, I went on a mission to Ukraine together with Wim van Velzen, as we often did in the countries of the former East block or those that belonged to the Soviet Union. Together with local democracy-building partners we organised a seminar on "politics and religion". What I recall from those days is the diversity of the Christian churches and the luxury and high quality of dinners they provided. It is not by chance that Ukraine is called the breadbasket of Europe! But at the beginning of the nineties the possibility of a European perspective for Ukraine was not yet considered. This would change dramatically in 2004.

The first political contacts were established by the European Democrat Students (EDS), the Youth of the European People's Party (YEPP), and the Konrad Adenauer Foundation in Kiev. In October 2003 my Secretary-General Antonio

López-Istúriz led a fact-finding mission to Ukraine. In the framework of this visit, Viktor Yushchenko accepted an invitation to address the EPP Congress of February 2004 in Brussels. There I had the first personal meeting with him, during which he applied for EPP observer status for his political alliance, Our Ukraine, a bloc of parties that later won the elections. In November 2004 I visited Kiev to support Yushchenko in his campaign for the presidency. Ten days later I learned that against all expectations he had lost the elections. It was clear that there had been foul play. As people started pouring into the streets, the Orange Revolution began. Yushchenko was joined in the revolution by another important personality, namely Yulia Tymoshenko. Yushchenko and his Our Ukraine movement could count on EPP support. In one of his letters he expressed that my support and support of the European People's Party were very important to him personally and well-timed.

In support of the Orange Revolution, our Prime Ministers at the EPP Summit of 16 December 2004 made the following statement: "The democratic forces spearheaded by the Our Ukraine coalition and led by presidential candidate Viktor Yushchenko, and the gradual convergence of political and civil society forces in the Our Ukraine coalition, inspired by EPP values, will offer a common European future to the Ukrainian people". A month later, the EPP Political Bureau approved unanimously the application of Our Ukraine for EPP observer-member status (28 January 2005). Earlier that month, Yushchenko had finally prevailed and was sworn in as the new President.

On 22 November 2006 I met for the first time with Yulia Tymoshenko at the EPP headquarters in Brussels and explained to her the history of our political family. With friends and colleagues we afterwards enjoyed dinner together at the Stanhope Hotel opposite the EPP's offices. Yulia had contacts with nearly all of the European political parties. She would only decide a few months later which European party she would adhere to. The new parliamentary elections of 30 September 2007 were crucial. Yulia invited me to address the congress of her Bloc Yulia Tymoshenko (BYuT) on 5 August. "The EPP supports the unity of the Orange forces", I said, "You need a stable government and not a mafia, capable of acting in a timely and efficient way – and committed to protecting democracy and the rule of law". The congress approved her proposal that her Batkivshchyna (Fatherland) Party would join the EPP. Two days later I was a guest at the congress of President Yushchenko's Our Ukraine–People's Self-Defense Bloc, where I made the same appeal for unity.

On election day, the Orange parties obtained a majority in Ukraine's parliament, the Verkhovna Rada. The President and his Prime Minister–elect were received as victors at the EPP Summit in Lisbon on 18 October 2007. They committed themselves to form an "Orange" government and thanked our leaders for their support during the campaign as well as their support for the recognition of the Holodomor, or forced famine. In 1932 and 1933 millions of innocent victims died of hunger in Ukraine because of the orders of Stalin's regime. The European Parliament on 23 October 2008 adopted by an overwhelming majority a resolution commemorating the victims of the Holodomor. Our EPP-ED Group was one of the initiators of the resolution and our Group Chair, Joseph Daul, stated: "We in Europe have a special responsibility to recall crimes such as this, so that these crimes can never happen again".

On 7 February 2008 we accepted unanimously Yulia Tymoshenko's Fatherland Party as an EPP observer member. Soon after, discord between the Orange parties re-emerged. On 13 March 2008 the EPP Summit expressed its wish that the Orange coalition government would continue its reform programme. Subsequently at the same venue, I had a private meeting with President Yushchenko. His entourage tried in vain to prevent our meeting because I made sure to point out that many key figures were destabilising the government and that the democratic forces of Ukraine could collapse. During the summer I tried to organise a committee of "wise men" that would include former Polish President Aleksander Kwaśniewski and the former Spanish and British Prime Ministers José María Aznar and Tony Blair, to mediate between the two Orange rivals. However, due to the lack of agreement on a suitable date, a meeting could not be convened in time, and the attempt, such as it was, failed ....

The Caucasian Conundrum

When on 7-8 August 2008 the conflict in Georgia broke out, relations with Ukraine became more complicated. On 9 August 2008 I expressed my deep concern over the violent fighting in South Ossetia and the attacks on Georgian civilian targets beyond South Ossetia carried out by the army of the Russian Federation. If the situation were not defused quickly, it had the potential to destabilise the entire Caucasus region. I called for the immediate cessation of hostilities by all parties and the swift resumption of negotiations in order to find a political solution to this crisis, and respect for the sovereignty and territorial integrity of Georgia.

Georgia's President Mikheil (Misha) Saakashvili and his United National Movement (UNM) had already in May 2007 applied for EPP membership. I took part in an EPP fact-finding mission organised by Corien Wortmann-Kool in early July 2007, when I had the opportunity to see first-hand this remote European country. When I asked in a meeting with Georgian NGO representatives if they considered Georgia to be a European country, one of them responded, "Georgia cannot afford not to be a European country. If Georgia ceases to be European then it will cease to be a country". After a string of political crises followed by the decisive presidential (5 January 2008) and parliamentary elections (21 May 2008) won by Saakashvili and his party, the UNM was finally granted observer-member status by the EPP Political Bureau on 16 September 2008.

Following the Georgia–Russia conflict, President Yushchenko's entourage spread the rumour that Yulia Tymoshenko chose to side with Moscow during the crisis and labelled her a "Kremlin conspirator". At the EPP Summit in Brussels on 15 October 2008 she refuted this criticism in the presence and with the full support of President Saakashvili. She reported on her string of clear statements since 20 August, when she emphasised that both she personally and the Ukrainian government support the territorial integrity of Georgia: "The territorial integrity of Georgia is a sacred issue both for Georgia and Ukraine". She could also refer to her contacts with UK Foreign Minister David Miliband on 27 August and Chancellor Angela Merkel on 29 August, when she gave her full support to the EU plan for resolving the conflict.

Meanwhile, President Yushchenko decided to dissolve the Verkhovna Rada again and hold early elections. But at the EPP Summit, government and party leaders strongly supported the revival and strengthening of the coalition of democratic pro-European forces; an effective and stable government in Ukraine is essential in the aftermath of the Georgian crisis as well as in the recent global financial crisis. New parliamentary elections, the third in merely three years, would undermine the stability of Ukraine as well as its European perspective.

I fear that there is still a long way to go in Ukraine. My long experience with the parties in Central and Eastern Europe has taught me that to create partnerships in politically unstable countries is very difficult and full of risks. But a pan-European party like the EPP that is eager to disseminate its programme and values across the continent cannot wait. To neglect such an important country like Ukraine is not an option for the EPP. We have always taken risks and that is why we are today the largest political movement in Europe. At the same time I am convinced that the engagement of pan-European parties in countries like Ukraine actually contributes to political stability and to democratic maturity. To sum up, I believe it is also appropriate here to include Yulia Tymoshenko's recent statement on how she sees the future relationship of Ukraine and Russia vis-à-vis the EU:

The existence or absence of a framework of cooperation often determines whether diplomatic disputes mutate into a crisis. Effective international frameworks, however, usually demand a grand bargain. The settlement between France and Germany over steel in 1952 resulted in a sharing of sovereignty over a vital state function. This agreement set a precedent for the entire European project. In 2009 the challenge for the EU is to begin to shape a bargain involving Russia and Ukraine (*The Economist*, 19 November 2008).

Test Case Turkey

I mentioned in Chapter II (A European People's Party) that the CDU/CSU wished to maintain structural relations with political parties outside the European Community that did not belong to the EPP. They decided in 1978 to create the European Democratic Union (EDU), an association of Christian Democrat, Conservative and other non-collectivist parties. Turkish centre-right parties like Mesut Yilmaz's Motherland Party and Tansu Çiller's True Path Party were welcomed with open arms into the EDU. Ironically, these parties had less trouble being accepted as members of the EDU than the Justice and Development Party (AKP) did when it was granted observer-member status in the EPP in January 2005.

During the EPP Summit in Brussels of June 1996, our heads of government declared their opposition to the accession of Turkey to the EU. This statement created great political upheaval and incomprehension in Ankara. Moreover, to make matters worse, Turkey's centre-right parties – Motherland and True Path – were erased from the political landscape in the 2002 elections. A totally new situation emerged because the AKP came to power; in its electoral debut, it obtained an absolute majority. Although Recep Tayyip Erdoğan was leader of the party, Abdullah Gül became Prime Minister.

During this time, the Konrad Adenauer Foundation office in Ankara issued a detailed analysis on the AKP and the social and political developments in Turkey. In the world after 9/11, I became more and more convinced of the importance of dialogue with a secular party like the AKP, which includes moderate, democratic Muslims in a country that makes a clear separation between religion and the state.

The first contacts between the AKP and EPP date back to late 2002. On 7 February 2003 I visited Istanbul together with Wim van Velzen and met with Erdoğan. At the invitation of AKP, EPP Vice-President Kostas Karamanlis addressed their first party congress on 12 October 2003. During this same month an informal fact-finding mission was undertaken by Wim van Velzen (CDA), Hartmut Nassauer (CDU) and Antonios Trakatellis (ND). The then Minister of Foreign Affairs, Abdullah Gül, addressed the EPP Congress in Brussels in February 2004. A few days later the widely commented visit to Turkey by Angela Merkel, leader of the opposition, together with Wolfgang Schäuble took place. Merkel met with Erdoğan in person. During my July 2004 visit to Ankara, Erdoğan announced that his party would apply for EPP membership.

After the European Commission's positive recommendation in its report of 6 October 2004 to start accession negotiations, the EPP Summit of 4 November gave its agreement to the AKP's EPP application, accepting it "initially" as an observer member rather than an associate member, according to the original request of Erdoğan. The EPP Summit also stated that Turkey's accession negotiations should be "open-ended". A day later I received Gül in my office to discuss and clarify the decisions of the EPP Summit. Following the decision of the Political Bureau of 28 January 2005, the AKP became an EPP observer member. On 22 March 2005 Prime Minister Erdoğan attended the EPP Summit in Meise.

No Muslim Democrats

The fact that the AKP became an observer member of the EPP was not a trivial event. It meant that the AKP accepted our statutes, including the preamble and our Basic Programme of Athens. But can this party also stand behind Christian Democratic principles? When I asked this question to Erdoğan he answered: "We have a strong affinity with the Christian Democrats. We share the same fundamental values and are, just as you, inspired by our religion". In their programme there is hardly any reference to Islam. "The AKP regards Ataturk's principles and reforms as the most important vehicle for raising the Turkish public above the level of contemporary civilisation and sees this as an element of social peace". Regarding fundamental rights and freedoms, their party considers religion as one of the most important institutions of humanity, and secularism as a prerequisite of democracy, and an assurance of the freedom of religion and conscience. It also rejects the interpretation and distortion of secularism as enmity against religion. Basically, secularism is a principle which allows people of all religions and beliefs to comfortably practice their religions, to be able to express their religious convictions and live accordingly, but which also allows people without beliefs to organise their lives along these lines. From this point of view, secularism is a principle of freedom and social peace. (AKP official website – English version).

In his letter of 8 September 2004 addressed to me, Prime Minister Erdoğan stated the following: "In the preamble to the statutes of the EPP there is a reference to the Christian view of mankind and the Christian Democratic concept of society. [...] The AKP will abide by this provision to the extent allowed by its own statute and in conformity with the Constitutional Treaty of the EU". Thereafter I put the following question to him: "Why does not AKP call itself Muslim Democrats as we in the EPP call ourselves Christian Democrats?" His answer was clear: "We can never do this. We can never use our religion in the name of the party because people, movements and parties make mistakes and this would have a devastating effect on religion". Instead, the AKP officially defines itself as "Conservative Democrat".

Together and With Us

So there are numerous reasons why close cooperation between the AKP and the EPP is justified. Our programme and views are closely linked. The European Liberals were ready to accept the AKP at any price, and for a moment such a development almost materialised when Erdoğan attended the summit of the Liberal government leaders in June 2004. Even the European Socialists started to take notice when the AKP achieved a landslide victory in the July 2007 elections, with 46.6% of the vote and 341 of the 550 parliamentary seats.

Nevertheless, it is in the AKP's interest to continue its cooperation with the largest European party that represents the centre-right. If Erdoğan seeks a breakthrough in Europe it will not be possible with the Liberals or the Socialists. By working with us, there is a greater chance that opposition to Turkey can be won over. This became clear again in the recent court case attempting to shut down the AKP. On 22 May 2008 I expressed my deep concern with the Constitutional Court's handling of this case:

In all the years of my European political engagement, this is the first time that I have ever witnessed the judiciary of an EU candidate country actually contemplating to outlaw its own governing party – a party that enjoys 50% of the support of the electorate; a party that belongs to the largest European political family. I am afraid that if the court follows through with this case, it will put the entire reform process of the country in jeopardy.

To my knowledge I was the only European party leader that publicly took a stand. After deliberating for three days, the court gave its verdict on 30 July 2008. A qualified majority of seven out of eleven votes is required to disband a political party. Six members of the Court voted in favour of disbanding the party, thus falling short of the required qualified majority by one vote. Four members voted to cut government funding for the party, while the Chief Justice rejected closing it down.

Indeed, some EPP member parties are still very uneasy about our cooperation with the AKP. This is definitely the case when the issue of Turkey is used for internal national political ends. I also have great difficulty understanding why François Bayrou has become a fervent opponent of Turkey's EU accession, even though his MEPs sit in the Liberal Group in the European Parliament (ALDE) – the group most outspoken in favour of Turkey's EU membership. The hostility of Valéry Giscard d'Estaing to Turkey's EU membership in an interview in *Le Monde* on 6 November 2002 also had a determining influence on French public opinion. The mental resistance of the French to Turkey is being reinforced by various opinion-makers. Yet if there is one country that had a crucial influence on the modernisation and westernisation of Turkey, it is France; in the nineteenth and the beginning of the twentieth centuries, French influence played a pivotal role in the birth of the modern and secular Turkey that we know today.

EC/EU membership was already put in perspective for the Turks in 1963. On 14 April 1987, when I was President of the European Council, they officially submitted their membership application. Today the EU of twenty-seven Member States is not in a position – neither institutionally nor economically – to integrate such a large country. But the current realities should not be viewed by Turkey as a rejection. The EPP is in favour of open-ended negotiations that will first lead to a "privileged partnership". It all depends on the results of this interim phase and the capacity of the EU to eventually integrate such a large country. If a privileged partnership is assessed positively and if the EU has sufficient financial means, arguments against full membership will disappear. Above all, we will need political leaders who are no longer the prisoners of their internal struggles but have the courage to defend the common good of Europe.

Moving to the World Stage

Our strong focus on Europe might almost make us forget that there is another world that is less occupied with internal vicissitudes and that struggles with problems of another magnitude. Christian Democracy is embedded in a larger global organisation called Christian Democrat International (CDI). The EUCD used to be the European regional organisation of the CDI, but when it merged with the EPP in 1999 the EPP took over this competence. But more importantly, together with the Chilean Presidents Eduardo Frei and Patricio Aylwin, the CDI has been the guiding force and often the political saviour of those in need.

Unfortunately, at the end of the 1990s the CDI was in a deep financial crisis. Upon findings from the European Court of Auditors, the EPP Group in the European Parliament cancelled its annual subsidy. This meant no less than sudden death for this long-standing international organisation. As the CDI was *de facto* broke, the Secretary-General of the EPP, Alejandro Agag, proposed to organise its liquidation. In view of his excellent connections in Latin America, I personally thought it would be a good idea if José María Aznar became CDI chair. Following his refusal to take on this task immediately, I proposed to chair it for an interim period.

Therefore, I became chair of the CDI at the congress in Santiago de Chile on 11 October 2000. In fact I was a sort of crisis manager who had to resolve material and financial problems. But I also made sure there was a new emphasis on the content by creating a new basic programme together with President Patricio Aylwin. This was approved on 20 September 2001 by the Leaders' Conference in Mexico City. On that day – as previously agreed – Aznar was elected chair. As well, the name of the international organisation was modified. With the same abbreviation, CDI, it was henceforth called Centrist Democrat International. In some Asian and African countries reference to religion is not allowed, and without the name change it would be impossible to extend the International further into Asia and Africa. Since the name change, many parties from Asia-Pacific and African regions have joined.

Nevertheless, there are still two multilateral international organisations of the centre-right that operate in parallel – as was the case some years ago in Europe with the EPP and the EDU, which was fortunately resolved with the merger of the EDU and the EPP in 2002. The International Democratic Union (IDU), founded in 1983, also assembles centre-right parties from around the world (including practically all EPP member parties). In fact, the initiative to form the IDU was taken by George H. W. Bush along with Helmut Kohl, Margaret Thatcher, Jacques Chirac and Alois Mock. Yet the decision to establish the IDU Secretariat in London led, inevitably, to tensions with the CDI (and the EPP). The Secretariat fell into the hands of unimportant staffers of the British Conservative party who were opposed to any form of cooperation with the CDI and dismissed altogether the EU perspective in international politics, i.e. the EPP.

Fortunately, change in the IDU came in the fall of 2004, when the IDU Executive Board and its chair, veteran Australian Prime Minister John Howard, decided to relocate the Secretariat from London to Oslo. The appointment of Høyre politician Eirik Moen to the position of IDU Secretary-General was welcome news, since Høyre is a very active member party in the EPP. As a result, the IDU became "mentally" closer to the EPP and the CDI. Soon after, I exchanged a number of letters with John Howard where together we paved the way for the EPP to become a regional union of the IDU. At the IDU executive meeting in Sydney Australia on 21 April 2008, the EPP was unanimously accepted as a regional union, alongside the EDU, which has remained as the nominal entity of the IDU for parties that are not members of the EPP (i.e. British Conservatives, the Czech ODS). The decision was confirmed at the IDU Party Leaders Conference in Paris in July 2008, where I was elected IDU Vice-Chair.

Nevertheless, even though the EPP now functions as a bridge between the IDU and the CDI, the majority of member parties in both internationals insist on a IDU-CDI fusion. I firmly believe that our Christian Democratic values would be better served worldwide if we finally managed to establish a strong, united international organisation of the centre-right. The impact of such an organisation would increase considerably; it would offer greater benefits to its members and would pose a serious challenge to the well-established Socialist International. Yet a serious initiative in this direction has failed to materialise. Many problems need to be addressed, such as the reservation of some Latin American CDI member parties about cooperating with the US Republicans.

Apart from our multilateral involvement in international politics, the EPP embarked in 2005 on an effort to establish its bilateral international contacts with parties and countries beyond the EU and, more importantly, beyond the European continent. Considering that the EU has worked hard in recent years to develop its international profile, it was only natural for the EPP to develop its own international profile in the same direction. In fact, our work reinforces the work of the European Commission and its External Relations Service, since our goal is to develop contacts with parties and governments around the world that are preferably like-minded and seek stronger relations with the European Union. With the amendment of the Statutes/Internal Regulations of the EPP in 2006, the position of Secretary of External Relations was established for the first time, and in September 2006 the Political Bureau elected Kostas Sasmatzoglou of Nea Demokratia to the position.

A characteristic example of our work has been our bilateral relationship with the US Republican Party. This relationship was made possible through the partnership that we established in 2005 with the International Republican Institute (IRI), a democracy-building organisation that functions as the informal international office of the Republican Party. Since then we have co-organised with the IRI many events with high profile speakers, including an event in Brussels on 29 April 2006 featuring IRI chair (and 2008 presidential candidate) Senator John McCain. Equally important was the EPP's participation at the Republican Convention in

St Paul, Minnesota (1–4 September 2008), where we were represented by a strong delegation and organised a successful EPP public relations event. In the aftermath of John McCain's defeat on 4 November 2008 the Republicans must embark on a phase of renewal for their party. Our cooperation with them is not only necessary, it is also a matter of political solidarity. I also hope that our intention to open an office in Washington DC – in order to promote more effectively the EPP's perspective on EU–US relations – will soon materialise.

The EPP's international activities have also expanded to Canada, where we have built strong contacts with Prime Minister Stephen Harper and the Conservative Party of Canada, as well as with the Liberal Party of Australia, the Liberal Democratic Party of Japan, the New Zealand National Party and others. Paradoxically for many, we have also established contacts with the Communist Party of China, not because we agree in ideological terms with them but because we need to communicate so that we can improve mutual understanding as well as criticise them on human rights issues and China's lack of democratic reforms.

Surprisingly, our new political innovation to develop political contacts on the international level has yet to be matched by any of the other European political parties.

Epilogue

Adam: Every Human Being and the Whole of Mankind

St Paul to the Corinthians:

Charity suffereth long, and is kind; charity envieth not; charity vaunteth not itself, is not puffed up,
Doth not behave itself unseemly, seeketh not her own, is not easily provoked, thinketh no evil;
Rejoiceth not in iniquity, but rejoiceth in the truth;
Beareth all things, believeth all things, hopeth all things, endureth all things.
(...)
For now we see through a glass, darkly; but then face to face: now I know in part; but then shall I know even as also I am known.
And now abideth faith, hope, charity, these three; but the greatest of these is charity.
(1 Corinthians 13, King James Version)

How can a politician live out St Paul's ideal as he depicted it in the letter to the Corinthians? For he is not talking about the love between a man and a woman, but rather about fraternal love among Christians in one of their first communities. In my early days in politics I was deeply affected by the French philosopher Paul Ricœur. I read and reread his *Essays*. In 1968, a collection of articles was published in the Netherlands as part of a volume entitled "Politics and Faith" (*Politiek en geloof, Essays van Paul Ricœur*, Utrecht: Uitgeverij Ambo, 1968). These texts helped me to better understand and embrace the ideals and virtues of St Paul and to strive for their realisation.

In this epilogue I reflect on my social and political stance. I attempt to decipher what lies behind my engagement and, once again, Paul Ricœur is my unrivalled "fellow traveller". His essay, "The *Socius* and the Neighbour" (1964), deals with the core of this entire issue. He stresses the historical scope of love and the great wealth concealed in the dialectic of the *socius* (fellow citizen) and the neighbour. "Sometimes the personal relationship to one's neighbour occurs *via* the relationship to the *socius*; sometimes it develops *in the margin* of social relationships; sometimes the two kinds of relationships are in *opposition* to one another". Referring to the parable of the Good Samaritan, he contends that the neighbour also signifies the double requirement of closeness and distance: "So the Samaritan was close because he came closer and distant because he was still the non-Jew who on a particular day picked up a stranger from the road".

"The neighbour" is the personal way in which I encounter another person, irrespective of any social mediation. It is an encounter, the point of which precludes any judgement based on history. The ultimate point of institutions – in other words, of such social mediation – is the service they can render to people. If no one benefits from them the institutions are pointless. But their significance remains hidden because no one can estimate their beneficial effects.

Love is not necessarily found where it is most obviously exhibited; it also exists hidden in the humble, abstract services of post offices and social security departments; very often it is the hidden point of social interaction. The Last Judgement means that we "will be judged" according to what we have done for individual people through the medium of the most abstract institutions and even without being aware of it. The servant love shown towards individual people will in the end be the overriding one. That is the very thing that remains *surprising*. For we do not know when we reach people. We think we exercise this direct love in the "immediate" relationships of person to person, but our love is often nothing other than exhibitionism; and we think that we do not reach anyone in the "more distant" relationships of work, politics, etc., but perhaps we are wrong in this too. The measure of human relationships seems to be: "Are we reaching individual people?" But we have neither the right nor the ability to employ this measure. ("Le *socius* et le prochain", in *Histoire et vérité*, Editions du Seuil, 1964, pp. 99–111, our translation).

In 1960, Paul Ricœur took up this train of thought again in "L'image de Dieu et l'épopée humaine". "One possible route is the one I took a few years ago in the essay 'The *Socius* and the Neighbour'", he says:

At that time I compared the "immediate" relationships of person to person – relationships with one's neighbour – to the "more distant" relationships conducted via institutions and social organs – relationships with the *socius*. Contemplating the image of God enables us to start out from what was, at that time, the destination, in other words the profound, hidden unity of all these relationships as expressed in the theology of love, which is at the same time a theology of history. The Church Fathers knew that man is both individual and collective. Man – that is, every human being *and* the whole of mankind. Some of them also knew that Adam means human being, *anthropos*. They were able to imagine a collective individual, an individual that is the equivalent of a race, a collective which branches off into individual thoughts, desires and feelings.

"They could still understand this paradox", says Ricœur, because they had preserved the historical and cosmic dimension of the image of God. It is obvious that this paradox surprises us, in these times: as never before, our human race threatens to break into two, between the immediate relationships of friendship, married life and private life on the one hand and the mediated relationships of economic, social and political life on the other. This dichotomy of the private and the public, which makes nonsense of both, is the very opposite of an anthropology which takes its initiative from meditation on the image of God ("L'image de Dieu et l'épopée humaine", in *Histoire et vérité*, Editions du Seuil, 1964, pp. 112–131).

What does this mean for us, we who are Christian Democrats? Europe is undergoing a period of profound change. Since the end of the Cold War, our countries and peoples have been struggling to adjust to the ever-changing circumstances in Europe and the globe. These new circumstances have presented Christian Democrats with new challenges but also with new opportunities. Long-standing European democracies are experiencing a deep crisis of values that is challenging the political system at its very foundations. Moreover, Europe's newly established democracies are still trying to deal with the growing pains of their political system while, at the same time, coping with EU accession and economic globalisation.

We are nonetheless aware of the dangers, temptations and the loss of direction resulting from the crisis of values that Europe is experiencing. We cannot provide simple solutions to overcome all these challenges, but we can act on the basis of fundamental values and principles that will point us in the right direction.

Against Ideological Temptations

Despite the collapse of Marxism–Leninism in Central and Eastern Europe, the end of ideological competition in Europe is far from over. Liberalism, for example, offers many advantages. Its prescription for how the market economy should function has made possible the development of living standards that other systems have not yet been able to achieve. Neo-liberalism ignores, however, the social dimensions of the free market economy by unilaterally stressing the individual efforts of each man and woman.

Ecology, which is compatible with Christian Democratic values, provides a positive contribution to resolving pressing challenges such as climate change and strives to bring out the best in every man and woman for advancing the quality of life. Within this ideology, however, lies the temptation to attribute absolute values to nature and the earth but to devalue all technological and economic progress.

Socialism is an ideology full of handicaps; either it uses the old model of class struggle and class opposition, a model which no longer satisfies its citizens' aspirations, or – as is the case in social democracy – it gives up the class fight but remains suspicious of civil society and intermediary bodies and gives too much priority to the regulatory activity of the state, thus all too often invading the social fabric.

Needless to say, the most dangerous ideology in Europe currently is the amalgam of populism and nationalism, or "populist nationalism". The feeling of patriotism and of belonging to a specific community is inherent in the existence of every human being and thus entirely legitimate. Populist nationalism, however, is not acceptable since it denies in absolute terms other values and responsibilities such as loyalty to the democratic state and the rights of minorities.

Europeans differ and are often contradictory in their perception of values. The growth of materialism, individualism and cynicism, together with the collapse of universal ideals on the one hand, and the drive towards resolving individual problems in society (single-issue politics) on the other, has created a complex social environment. But at the very least, this new environment has also created a renewed aspiration to lead one's life on the basis of values such as responsibility, loyalty and a concern for security.

Hence, now more than ever, Christian Democracy and all like-minded political movements must identify the core concerns of European societies and offer modern, tangible solutions and a vision for a prosperous future.

Overcoming New Challenges

European society sees itself confronted by multiple contradictory developments that jeopardise its internal cohesion.

Negative demographic trends, for example, pose a serious challenge to Europe and will continue to dominate our societies over the coming decades. At the same time, significant migratory movements from the East and South have compounded the pressure on Europe.

Scientific knowledge is a potential source for an improvement in well-being and for the fair distribution of prosperity. Technological innovations will help improve the quality of tomorrow's society. Nevertheless, technological and scientific developments should not mislead us or cause us to lose respect for nature.

The flow of meaningless information, especially when distorted by the media, may result in the total disappearance of responsibility and the introduction of irrationality into politics. It can create a climate of instability and confusion as well as lead to apathy and the impression that there are quick and easy solutions to all problems.

Christian Democrats do not claim to have all the solutions to these challenges. Without losing sight of the dangers that exist, however, I believe that we should also seize new opportunities. For example, we should take the lead in spreading and strengthening democracy around the world and, in effect, promote European values and norms around the world. The Christian Democratic model of linking the social market economy with democracy has already proven to be very successful in Central and Eastern Europe and offers great possibilities for many countries in transition around the world.

Fundamental Values

The European People's Party has made these ideals explicit and concrete within the context of our party activities and the European integration process. Those things that fundamentally unite us, notwithstanding our specific origins and different political cultures and religious traditions, were mutually established in our Basic Programme at our historic congress in Athens in November 1992. This document, which was drafted in the aftermath of the Cold War, was intended to provide direction for the new party in a period of great uncertainty. It puts into words the cornerstone of our political project. Not surprisingly, the relevance and expressive power of this document continues to be stronger than ever.

Christian Democracy should not be taken for granted. It is constructed on a carefully crafted set of values that are rooted in the essence of every human being. It affirms the dignity of the human being and regards humanity as the subject and not the object of history. On the basis of Judaeo-Christian values, it holds that each person is unique, irreplaceable, totally irreducible, free by nature and open to transcendence.

Because they are free, responsible and interdependent, people must take part in the construction of society. For many, what lies behind this commitment is the belief that we are called to contribute to God's work of creation and freedom. Freedom is inherent in the essential nature of humanity. It means that every individual has the right and the duty to be fully responsible for himself or herself, for his or her acts, and to share responsibility for neighbours and all living creatures.

Truth is transcendent and, as such, is not entirely accessible to humanity. Freedom affirms that humanity is by its nature fallible. Consequently, Christian Democrats acknowledge that it is impossible for anyone to conceive or even to construct a perfect society, free of all pain or conflict. At the same time, Christian Democrats reject any form of totalitarianism based on such an aspiration.

Every woman and every man is responsible for constantly improving society on the basis of core values and regulating principles that are established in common. By applying these values and principles, we are able to prevent, manage and settle peacefully the differences between humans and the challenges facing humankind.

Freedom and Responsibility

Our thought and political action are based on fundamental, interdependent, and equally important universal values: freedom and responsibility, fundamental equality, justice and solidarity.

Genuine freedom means autonomy and responsibility, not irresponsible independence. Everyone shares responsibility for the created world. Future generations must also be able to live in harmony with a natural environment in which each human being is an essential link in the chain.

Equality

All human beings are equal because they are endowed with the same dignity. Notwithstanding their differences in terms of talents and abilities, all humans must be free to achieve their personal development regardless of origin, sex, age, race, nationality, religion, conviction, social status or state of health.

Justice

Justice means that the conditions in which individuals can exercise their freedom must be guaranteed at all times. One dimension of justice is respect for the rule of law. Laws are constantly evolving, but they must always be freely accepted by people.

Laws must evolve on the basis of respect for the fundamental and inalienable rights of all humans, as defined in the 1948 Universal Declaration of Human Rights and the 1950 European Convention on Human Rights and Fundamental Freedoms.

Justice cannot be arbitrary; it cannot be perceived as the tyranny of the majority. It respects the minority and no majority can deny the minority the free exercise of its rights.

Solidarity

Justice, fundamental equality for all and the dignity of every individual are inseparable. Solidarity means an awareness of the interdependence and interrelatedness of individuals and their communities. It means above all protecting the weakest in our society and around the world.

The Christian Democratic concept of the person rejects selfish individualism and collectivism as reductive. It rejects extremes and exclusion – it advocates dialogue, tolerance and sharing.

Respect for the Created World

Christian Democracy opposes the unjust exploitation of the earth, the lack of respect for the self-regenerating potential of nature. It calls for the management of the earth with a view to satisfying the needs of all and improving the living conditions and quality of life of everyone, while also ensuring lasting development compatible with the protection of the interests of future generations.

Respect for the created world means that responsible management of the biosphere and humankind's common heritage is necessary and essential for the harmonious development of every living and future human being.

Subsidiarity

Subsidiarity means that power must be exercised with a view to solidarity, effectiveness and the participation of citizens; in other words, where it is most effective and closest to the individual. Tasks that can be performed at a lower level must not be transferred to a higher level.

Subsidiarity is based on the premise that society can be freely constructed. Public authorities must respect human rights and fundamental freedoms, recognise the autonomy of civil society and not take the place of social and private initiative. The activities of non-governmental organisations and the creation of associations of every form must be encouraged.

Democracy

Christian Democrats consider democracy as vital to the development of individuals. The participation of each person in public life and in decisions is essential for democracy. This commitment implies the strengthening of the constitutional state.

Applying subsidiarity, international partnership and participation by all, especially through free elections, enables each person to achieve respect for others and to work towards the resolution of conflicts.

The limits imposed by subsidiarity contribute to the separation of powers and prevent their concentration. Every authority is, in effect, at the service of the individual. No state, therefore, can use the pretext of respect for its sovereignty in order to violate the rights and freedoms of persons or communities.

The raison d'être of the sovereignty of states is to enable them to work freely and as best they can in order to ensure the well-being and development of their people and to defend international law. This also means that states must share their sovereignty in supranational and international organisations.

Inter-religious and Cultural Understanding

Given that since 1992 we have welcomed like-minded political parties, we have also accepted and created religious pluralism within the EPP. Its membership now includes Catholics, Protestants, Orthodox Christians, Jews, Anglicans and even Muslims. We believe that this plurality of religious backgrounds does not pose a threat to the fundamental and all-encompassing values of Christian Democracy. On the contrary, it is precisely our concept of humanity that unites us.

For this reason, prospective member parties of the EPP must accept, in addition to the party statutes, our Christian-inspired Basic Programme of Athens. A federated European political structure, the principle of personalism, the importance of the individual in a pluralistic society, the principle of subsidiarity and a social market economy are a few of the principal pillars of the EPP. Most of these pillars are part of the fundamental values of Christian Democracy and have to be agreed to by any party wishing to join the EPP family.

I believe that the Basic Programme of Athens and its religious pluralism made religious dialogue between the EPP and various Christian denominations a great success. The contribution of these dialogues to promoting inter-religious and cultural understanding as well as to resolving societal problems is enormous. The dialogues have, moreover, given the EPP a better knowledge of religious issues and reinforced solidarity by creating a more harmonious society. The relevance of these dialogues is amplified by the fact that immigration has changed the demographics of Europe.

It is now over ten years since the first dialogue between the EPP and the Orthodox Church took place. These rich exchanges have been an instrument of progress, having served as a tool for bringing together peoples in South-Eastern Europe after the fall of Communism and for reconciling them with the European Union.

An Appeal to Values

Europe is no longer what it was fifty years ago. Wars between nations have given way to increasing cooperation and integration among countries. Nation states have started operating harmoniously within transnational, federated political and economic systems. Multiculturalism and inter-religious tolerance are increasingly present. In keeping with these changes, the Christian Democratic movement has evolved and adapted itself to the social and demographic shifts within Europe. This development has been relevant not only for confronting diverse challenges; given the outspoken pro-European attitude and experience of Christian Democracy, the Party is currently in a perfect position to play an essential role in almost all phases of the European integration process.

The recent institutional crisis following the rejection in France and the Netherlands of the European Constitution and the negative Irish vote on the Lisbon Treaty are clear reminders of the importance of building trust and restoring confidence in our common beliefs and vision. Instrumental in this regard are the convictions that inspire our values and our political action. Although the Christian faith is no longer the only source of inspiration, its presence is ever more essential in giving us the necessary tools to cope with contemporary challenges in Europe.

The European People's Party wants to help build a world that is based on freedom and solidarity, in which every man and every woman is viewed as a human being in all his or her fullness and complexity.

We stress the need to distinguish between the roles of church and state in society, between religion and politics. However, we reaffirm the link that exists between, on the one hand, Christian values based on the Gospel and the Christian cultural heritage and, on the other hand, the democratic ideals of freedom, the fundamental equality between humans, social justice and solidarity.

The European People's Party has always been and will always remain a political party of values. We derive our strength and our inspiration from constant reference to these values. If the EPP were to forget, neglect or dilute its values, it would be merely an instrument of power, without soul or future, while also forfeiting its universal and original message, which is based on a comprehensive awareness of the irreducible complexity of the human being and of life in society.

Bibliography

De Bijbel, Em. Querido's Uitgeverij B.V. & Uitgeverij Jongbloed, Haarlem, 2004
1931 Quadragesimo Anno encyclical: Subsidiarii officii principium, Catholic Church. Pope (1922–1939), Bib ID 150102
Bowie RR, Friedrich CJ (1954) Studies in federalism, Bib ID2169244. Little Brown, Boston
Cardinal Newman JH (1988) In: George Appleton (ed) The oxford book of prayer. Oxford University Press, Oxford
Daničić M, Mirosavoc T, Prestel B, Spahić T (2003) Good governance in Bosnia and Herzegovina. Felix, Holzkirchen, Obb. ISBN 9783927983526
Gaddis JL (2006) The cold war. Allen Lane (Penguin), London, ISBN-10: 0713999128
Magris C (1986) Danubio. Garzanti, Milan
Márai S (1993) Les Confessions d'un Bourgeois. Albin Michel, Paris
Martens W (1994) L'une & l'autre Europe. Éditions Racine, Brussels
May K (2005) Durch das Land der Skipetaren. Kreutzfeldt Electronic Publishing, Germany, ISBN: 3-89856-031-7
Musil R (2004) Der Mann ohne Eigenschaften. Der Hörverlag, München. ISBN 3899404165
Papini R (1996) The Christian Democrat International. Rowman and Littlefield, Lanham, ISBN 0847682994
Papini R (1995) Il Coraggio Della Democrazia: Sturzo E L'Internazionale Popolare Tra Le Due Guerre. Studium, Rome. ISBN 8838237417
Rakowski M (1995) Es begann in Polen Der Anfang vom Ende des Ostblocks. Hoffman und Campe, Hamburg
Ricœur P (1991) The socius and the neighbour. In Ricœur P, Valdés MJ (eds) Reflection and imagination. University of Toronto Press, Toronto, ISBN 0802058809
Ricœur P (1964) Histoire et vérité. Editions du Seuil, Paris
Ricœur P (1992) L'image de Dieu et l'épopée humaine. Northwestern University Press, Evanston. ISBN ISBN: 0810105985
Ricœur P (1968) Politiek en Geloof ("Politics and Faith"). Utrecht, Amboboeken
Van Der Velden B (2005) De Europese onmacht (The Powerlessness of Europe). J.M. Meulenhoff bv, Amsterdam, ISBN: 9029075651
Van Wilderode A (1947) Najaar van Hellas. Gebrocheerd, Antwerpen
Velestinlis-Fereos R (1998) Human Rights: Hellenic Declaration of 1797. Ant Publications, Athens. ISBN 9602329440

Weber M (1919) Politics as a vocation (Politik als Beruf). Duncker and Humblodt, Munich

Multatuli (2004) Gutenberg Ebook of Max Havelaar, retrieved 10 February 2004, http://www.gutenberg.org/files/11024/11024-8.txt

1948 Universal Declaration of Human Rights, Adopted and proclaimed by General Assembly resolution 217 A (III) of 10 December 1948, U.N. Doc A/810 at 71

Declaration of the Rights of the Child, Proclaimed by General Assembly resolution 1386(XIV) of 20 November 1959, G.A. res. 1386 (XIV), 14 U.N. GAOR Supp. (No. 16) at 19, U.N. Doc. A/4354.

European Convention for Protection of Human Rights and Fundamental Freedoms, Council of Europe, Rome, 4.XI.1950, Entry into force: 18. May 1954; Protocols: Protocol No. 2 (ETS No. 44) of 6 May 1963 and amended by Protocol No. 3 (ETS No. 45) of 6 May 1963, Protocol No. 5 (ETS No. 55) of 20 January 1966 and Protocol No. 8 (ETS No. 118) of 19 March 1985

Treaty on European Union (Maastricht Treaty), Official Journal C 191, 29 July 1992

The European Social Chapter of the 1992 Maastricht Treaty, Official Journal C 191, 29 July 1992; Integrated in Amsterdam Treaty, Official Journal C 340, 10 November 1997

Treaty establishing the European Economic Community (EEC) and the European Atomic Energy Community (EAEC) 25.03.1957, entry into force 1 January 1958. http://eur-lex.europa.eu/en/index.htm

Treaty of Amsterdam amending the Treaty on European Union the treaties establishing the European Communities and certain related acts signed on 2 October 1997, Luxembourg: Office for Official Publications of the European Communities, 1997 ISBN 92-828-1652-4; Official Journal C 340, 10 November 1997

Single European Act (1986), Official Journal L 169 of 29 June 1987

Treaty of Nice Official Journal C 80 of 10 March 2001

AGENDA 2000, For a stronger and wider Union, Document drawn up on the basis of COM (97) 2000 Final, Bulletin of the European Union Supplement 5/97, 15 July 1997 European Commission, Luxembourg: Office for Official Publications of the European Communities, ISBN 92-828-1034-8 1997

Treaty of Lisbon amending the Treaty on European Union and the Treaty establishing the European Community, signed at Lisbon, 13 December 2007, Official Journal C 306, 17 December 2007

European Constitution, ISSN 1725-2423, Official Journal of the European Union, C 310 Volume 47, 16 December 2004

The Belgian Constitution, consolidated text, Belgisch Staatsblad, 17 February 1994

1980 State Reform, the Special Act of 8 August 1980 on the Reform of the Institutions, Belgisch Staatsblad, 15 August 1980

Law of 3 April 1990 on the termination of pregnancy, as amended by Sections 348, 350, 351, and 352 of the Penal Code of 1867, and repealing Section 353 of the said Code. POPLINE Document Number: 064462, Belgisch Staatsblad, 5 April 1990, pp

The General Framework Agreement for Peace in Bosnia and Herzegovina, reached at Wright-Patterson Air Force Base near Dayton, Ohio 21 November 1995, formally signed in Paris 14 December, 1995

Statutes of the EPP Group, adopted at the Constituent meeting of the European People's Party, Luxembourg, 8 July 1976

A federal and democratic constitution for the European Union, adopted at VIII EPP Congress, Dublin, 15-16 November 1990

Basic Programme of Athens 1992, adopted at IX EPP Congress, Athens, 11-13 November 1992

Europe 2000: Unity in Diversity, adopted at X EPP congress in Brussels, 8-10 December 1993

EPP- Force of the Union, adopted at XI EPP Congress in Madrid, 6-7 November 1995

On the way to the 21st century and the 1999-2004 Action Programme, adopted at XIII Congress in Brussels, 4-6 February 1999

The Western Balkan Democracy Initiative Report, Nea Demokratia, Athens September 2002

Index of Names

A

Abecassis, Snu, 136
Adenauer, Konrad, 2, 23, 41, 183, 213, 217
af Ugglas, Margaretha, 133
Agag Longo, Alejandro, 117, 220
Albert II, King, 46, 87
Alliot-Marie, Michèle, 162
Amaro da Costa, Adelino, 136
Amato, Giuliano, 169, 171
Anastasides, Nikos, 200
Andreotti, Giulio, 92, 96, 101, 103–106, 109
Andriessen, Frans, 187
Antall, József, 192
Arzalluz, Xavier, 113, 115, 116, 147
Aylwin, Patricio, 220
Aznar, José María, 113–118, 120, 127, 137, 140, 142, 148, 154, 158, 165, 167, 179, 183, 215, 220
Azzolini, Claudio, 141

B

Balkenende, Jan-Peter, 173, 179, 188
Bakoyannis, Dora, 124, 125
Barbi, Paolo, 130
Barnier, Michel, 170, 177
Barre, Raymond, 60
Barroso, José Manuel Durão, 137, 174–179, 199
Barrot, Jacques, 162
Bartholomew I, Ecumenical Patriarch, 150
Băsescu, Traian, 203
Baudis, Dominique, 162
Baudouin, King, 16–20, 33, 62, 65, 66, 78, 79, 80, 82, 83, 86, 87, 157, 191
Bayrou, François, 147–148, 161, 162, 166, 180, 219

Bech, Joseph, 186
Bendtsen, Bendt, 134
Berisha, Sali, 209, 210
Berlusconi, Silvio, 116, 132, 140, 141, 144–146, 158, 170, 174, 176, 177, 179, 180, 205
Bildt, Carl, 133, 142
Bizimungu, Casimir, 84
Blair, Tony, 56, 57, 176, 177, 205, 206, 215
Boc, Emil, 203
Böge, Reimer, 152
Bondevik, Kjell Magne, 134
Borisov, Boyko, 205
Boutros-Ghali, Boutros, 87
Bowie, Robert R., 13
Brandt, Willy, 210
Brittan, Leon, 126, 132
Brok, Elmar, 169, 170
Bruton, John, 111, 112, 126, 147, 170
Bukman, Piet, 187
Busek, Erhard, 164
Bush, George H.W., 93, 100, 221
Bush, George W., 82, 83, 158, 210
Buttiglione, Rocco, 139, 179
Buzek, Jerzy, 198

C

Cameron, David, 122, 160, 195
Carlsson, Gunilla, 133
Carrington, Peter Lord, 76
Castagnetti, Pierluigi, 140, 141
Cavaco Silva, Anibal, 104, 132, 136, 137
Chamberlain, Neville, 194
Chernenko, Konstantin, 76
Cheysson, Claude, 92

Chirac, Jacques, 39, 93, 142, 162, 171, 173–177, 186, 221
Christofias, Dimitris, 200
Churchill, Winston, 93
Çiller, Tansu Penbe, 217
Ciorbea, Victor, 202, 203
Ciuhandu, Gheorghe, 203
Claes, Willy, 44, 72
Clerides, Glafkos, 200
Cockfield, Arthur, 92
Collard, Leo, 26, 27, 30
Colombo, Emilio, 190
Constantinescu, Emil, 202
Cools, André, 44
Coposu, Corneliu, 201, 202
Cossiga, Francesco, 90
Coty, René, 11
Cox, Pat, 148
Craxi, Bettino, 92
Cresson, Edith, 153
Čunek, Jiří, 194

D
Daladier, Édouard, 194
Daul, Joseph, 215
Dávid, Ibolya, 193
David, Mario, 137, 144, 175
Davignon, Steve, 92
de Boer, Hans, 187
de Brouwer, Alain, 113
de Clercq, Willy, 54, 56–58, 95
De Gaulle, Charles, 23
de Larosière, Jacques, 53, 56, 57
de Palacio, Ana, 170
de Palacio, Loyola, 118
de Rougemont, Denis, 175
de Schryver, August-Edmond, 36
de Vries, Gijs, 148
Dehaene, Jean-Luc, 24, 49, 54, 62, 95, 96, 107, 110, 118, 126, 127, 142, 154, 156, 169–171
Dekker, Eduard Douwes, 187
Del Ponte, Carla, 205, 206
Deleeck, Herman, 30
Delors, Jacques, 91, 92, 94–98, 100, 102, 103, 126, 127, 131, 132, 151, 154, 171, 181, 183
Dimitrakopoulos, Giorgos, 182
Dinkić, Mladjan, 211
Djindjić, Zoran, 210
dos Santos, José Eduardo, 87

Dubček, Alexander, 194
Dukes, Alan, 110
Duran i Leida, Joseph, 147
Dzurinda, Mikuláš, 196, 205

E
Elles, James, 183
Erdoğan, Recep Tayyip, 217–219
Espersen, Lene, 134
Eurlings, Camiel, 187
Eyskens, Gaston, 19, 23, 25, 29, 34
Eyskens, Mark, 53, 54, 82, 84

F
Fenech Adami, Eddie, 200, 201
Fischler, Franz, 133, 152
Fitzgerald, Garret, 111
Fontaine, Nicole, 162
Fraga Iribarne, Manuel, 114
Franco, Francisco, 114
Frattini, Franco, 145, 179
Frei, Eduardo, 220
Friedrich, Carl J., 13
Freitas do Amaral, Diogo de, 136
Frost, David, 93

G
Gaddis, John Lewis, 71
Galeote, Gerardo, 141, 165
Gasperi, Alcide de, 1, 41, 113
Genscher, Hans-Dietrich, 135
Georgievski, Ljubčo, 212, 213
Gere, Richard, 150
Gil-Robles, José María Jr., 126
Gil-Robles, José María Sr., 114, 149
Giménez, Ruiz, 114
Giscard d'Estaing, Valéry, 89, 162, 169, 170, 219
Gol, Jean, 54
Gonzalez, Felipe, 96, 101, 104, 117, 132, 151
Gonzi, Lawrence, 201, 205
Gorbachov, Mikhail, 71, 78, 93, 100, 102, 151
Gotovina, Ante, 205, 206
Green, Pauline, 145, 148, 149, 153
Gromyko, Andrei, 76
Gruevski, Nikola, 213
Guardini, Romano, 2
Guigou, Elisabeth, 99, 107

Gül, Abdullah, 217
Gyurcsány, Ferenc, 193

H
Habyarimana, Juvénal, 82–86
Hague, William, 158, 159
Haider, Jorg, 164, 166
Hänsch, Klaus, 149
Harmel, Pierre, 71
Harper, Stephen, 222
Hassan II, King, 79
Haughey, Charles J., 104, 111
Havel, Václav, 194
Heath, Edward, 119
Helgers, Hans, 147
Hennicot-Schoepges, Erna, 147
Hennig, Ottfried, 142
Herman, Fernand, 126
Herzog, Roman, 168
Hintze, Peter, 174
Hitler, Adolf, 194
Houben, Robert, 31
Houthuys, Jef, 59, 61
Howard, John, 221
Howard, Michael, 159, 176
Howe, Geoffrey, 95
Hoxha, Enver, 209
Hussein, Saddam, 80, 102

I
Iliescu, Ion, 202, 203
Itälä, Ville, 134

J
Janša, Janez, 205, 207
Jansen, Thomas, 109, 111, 122, 126, 133, 137
Jaruzelski, Wojciech Witold, 196, 197
Jenkins, Roy, 90
John Paul II, Pope, 20, 150, 196
Juncker, Jean-Claude, 142, 148, 154, 173, 174, 177, 179, 186, 205
Juppé, Alain, 142, 162
Jurgens, Erik, 188

K
Kádár, János, 191
Kaczyński, Jarosław, 197, 198
Kaczyński, Lech, 198
Kagame, Paul, 81

Kalousek, Miroslav, 194
Kalvītis, Aigars, 199, 205
Karamanlis, Konstantinos, 124
Karamanlis, Kostas, 125, 174, 176, 179, 205, 212, 217
Katainen, Jyrki, 134, 135
Kegel, Sandra, 9
Khol, Andreas, 163
Kiszczak, Czesław, 197
Klaus, Václav, 194, 195
Klepsch, Egon, 112, 120, 121, 130
Kohl, Helmut, 39, 92, 96, 97, 104, 109, 120, 128, 130, 139, 146, 151, 170, 173, 184, 185, 190, 192, 221
Kok, Wim, 154
Kostov, Ivan, 204
Koštunica, Vojislav, 210
Kovács, László, 179
Kruisinga, Roelof, 136, 187
Kubilius, Andrius, 199
Kwaśniewski, Aleksander, 215

L
Laar, Mart, 199
Lamassoure, Alain, 162, 170
Lamy, Pascal, 94
Landsbergis, Vytautas, 199
Lecanuet, Jean, 161
Lefèvre, Theo, 15, 23, 36
Leinen, Jo, 182
Leopold III, King, 16, 23, 64
Leterme, Yves, 46
Liikanen, Erkki, 153
López-Istúriz, Antonio, 117, 213–214
Lubbers, Ruud, 75, 95, 99–101, 104–107, 121, 126, 127, 188
Lücker, Hans-August, 35–37, 39–42, 184
Lux, Josef, 194

M
Magris, Claudio, 191
Major, John, 103, 104, 120, 121, 126, 127, 158
Maniu, Iuliu, 201
Márai, Sándor, 6
Marcovici, Arlette, 201
Marin, Manuel, 153
Marini, Franco, 147
Maritain, Jacques, 1, 2

Martens, Anne, 4, 7
Martens, Emma, 16
Martens, Jacob, 1
Martens, Chris, 4, 7
Martens, Sarah, 4, 8
Martens, Simon, 4, 8
Martens, Sophie, 4, 8
Martens, Wilfried, 16, 30, 174
Matsis, Yiannakis, 200
Mauroy, Pierre, 91
Maystadt, Philippe, 62, 147
Mazowiecki, Tadeusz, 197
McCain, John, 222
McFarlane, Robert, 74
Mečiar, Vladimir, 195
Méhaignerie, Pierre, 161, 162
Méndez de Vigo, Iñigo, 170, 171
Merkel, Angela, 169, 175–179, 184–186, 216, 217
Michelis, Gianni de, 106
Mikhailova, Nadezhda, 204
Miliband, David, 216
Milošević, Slobodan, 210
Mitsotakis, Constantin, 104, 121, 123, 124
Mitterrand, François, 8
Mobutu, Joseph-Désiré, 78
Mock, Aloïs, 163, 164, 221
Moen, Eirik, 221
Monnet, Jean, 1, 41, 93
Monsengwo, Monseigneur Laurent, 81
Monteiro, Manuel, 136
Monti, Mario, 132, 145
Moro, Aldo, 139
Mounier, Emmanuel, 1, 2
Museveni, Yoweri, 83
Mussolini, Benito, 113, 194

N
Nagy, Imre, 191, 192
Nakasone, Yasuhiro, 94
Nassauer, Hartmut, 165, 217
Năstase, Adrian, 203
Niinistö, Sauli, 164
Novak, Ljudmila, 207
Nothomb, Charles-Ferdinand, 36, 42, 54, 135

O
Orbán, Viktor, 179, 192, 193, 211
Oreja, Marcelino, 115, 132, 167

P
Papini, Roberto, 113
Parts, Juhan, 199
Pasty, Jean-Claude, 141, 148
Patten, Chris, 120, 176, 208
Pawlak, Waldemar, 198
Pedersen, Niels, 134
Peterle, Alojz, 112, 170, 207
Petersen, Jan, 134
Petre Miluţ, Marian, 203
Piebalgs, Andris, 199
Pires, Francisco Lucas, 136
Pofalla, Ronald, 185
Poher, Alain, 161
Pöhl, Karl Otto, 58
Pollo, Genc, 210
Popescu-Tăriceanu, Călin, 203
Pöttering, Hans-Gert, 138, 141, 144, 157, 158, 160, 176, 177, 179
Prestel, Bernhard, 208
Prodi, Romano, 140, 145, 146, 153, 154, 173, 180
Prout, Christopher, 112, 120, 122
Pujol, Jordi, 114

R
Raffarin, Jean-Pierre, 163, 170, 177, 179
Rajoy, Mariano, 118, 186
Rakowski, Mieczyslaw, 197
Reagan, Ronald, 56, 71, 73–75, 78, 93
Reding, Viviane, 151
Reinfeldt, Fredrik, 135
Repše, Einars, 199
Ricœur, Paul, 1, 2, 28
Rinsche, Günther, 131, 151
Roman, Petre, 203
Rømer, Harold, 120, 134
Rupérez, Javier, 114, 115
Ruys, Manu, 30

S
Sá Carneiro, Francisco, 135, 136
Saakashvili, Mikheil, 215, 216
Sampaio, Jorge, 151
Sanader, Ivo, 205–207
Santer, Jacques, 36, 96, 101, 104, 110, 120, 121, 127–154, 186
Sarkozy, Nicolas, 162, 163, 186, 210

Sasmatzoglou, Kostas, 221
Savimbi, Jonas Malheiro, 87
Schäuble, Wolfgang, 158, 169, 217
Scheler, Max, 2
Schleicher, Ursula, 182
Schlüter, Poul, 104, 134
Schmidt, Helmut, 60, 68–70, 72, 74, 89–92
Schröder, Gerhard, 153, 154, 173–177
Schultz, George, 74
Schuman, Robert, 1, 41, 150, 161, 166, 180, 191
Schüssel, Wolfgang, 164, 166, 167, 175, 179, 205
Séguin, Philippe, 141, 162
Simeon Borissov Sakskoburggotski-Simeon II of Bulgaria, 204
Simonet, Henri, 72
Slota, Ján, 196
Smet, Miet, 155, 156
Smith, Ian Duncan, 159
Sousa, Marcelo Rebelo de, 137
Sramek, Jan, 194
Steenkamp, Piet, 187
Stoiber, Edmund, 158, 179
Sturzo, Luigi, 113, 114, 139
Sutherland, Peter, 126
Svensson, Alf, 133
Svoboda, Cyril, 194
Swaelen, Frank, 136

T
Tadić, Boris, 211
Thatcher, Margaret, 39, 60, 61, 75, 76, 89, 91–95, 97–101, 103, 105, 113, 119, 120, 127, 221
Thorn, Gaston, 90–92
Tindemans, Leo, 33–36, 41–44, 52–54, 73–78, 90, 110, 126, 130–132, 155
Tito [Josip Broz], 210
Topolanek, Mirek, 194, 195
Trakatellis, Antonios, 217
Tshisekedi, Etienne, 81
Tudjman, Franjo, 206
Twagiramungu, Faustin, 85
Tymoshenko, Yulia, 214–216

U
Uwilingivmana, Agathe, 85, 86

V
Van Acker, Achiel, 17
van Agt, Dries, 105, 188
van Buitenen, Paul, 152
van den Broek, Hans, 75, 104, 132
van der Velden, Ben, 154
van der Rohe, Ludwig Mies, 195
van Gennip, Jos, 123, 187
van Miert, Karel, 72, 95, 96, 149
Van Peel, Marc, 147
Van Peteghem, Leonce, 32
van Velzen, Wim, 165, 187, 190, 191, 213, 217
van Wilderode, Anton, 125
van Ypersele, Jacques, 56, 57
Vanden Boeynants, Paul, 44
Vanhanen, Matti, 134
Vassiliou, George, 200
Veil, Simone, 151
Velestinlis-Feraios, Rigas, 124
Veltroni, Walter, 181
Verhofstadt, Guy, 46, 60–62, 169, 173–177
Verplaetse, Fons, 56
Vittorino, António, 174, 175, 177
von Clausewitz, Carl, 180
von Habsburg, Otto, 151
von Hassel, Kai-Uwe, 39

W
Waigel, Theo, 142
Wałęsa, Lech, 101, 196, 197
Weber, Max, 28
Weinberger, Casper, 74
Welle, Klaus, 122, 126, 142, 160
Werner, Pierre, 57, 58, 186
Wintoniak, Alexis, 164
Wortmann-Kool, Corien, 187, 215

Y
Yilmaz, Mesut, 217
Yushchenko, Viktor, 214–216

Z
Zalm, Gerrit, 188
Zhivkov, Todor, 204

Annex

History and Chronology of the European People's Party

Brief History

Political formations of the centre-right can be traced back to the early 1920s. Unlike the case of the Socialists, Christian Democratic and Conservative pan-European cooperation was the child of national parties and derived from a federal tradition.

The experiences of the First World War and the threat of fascism led to the conviction among leaders that overcoming nationalism was the precondition for preserving peace. The first attempt at cooperation between like-minded Christian Democrats was made in 1926, when the International Secretariat of Democratic Parties of Christian Inspiration (Secrétariat International des Partis Démocratiques d'Inspiration Chrétienne, SIPDIC) was founded. However, fascism (National Socialism) increased tensions between governments, and the spirit of revenge and the dictators' obsession with power all eventually brought to an end cooperation among the Christian Social Democratic parties, and led finally to the outbreak of the Second World War.

The lessons and experiences of cooperation between 1925 and 1939 were key when leaders of the re-established or newly founded Christian Democratic parties in Europe formed the Nouvelles Équipes Internationales (NEI) in 1946. The ecumenical elements were decisive: reconstruction and reconciliation were born amidst the ruins of the national states, as was the vision of a united continent in the future.

Christian Democratic parties were banned in Central and Eastern Europe once communist rule was imposed. In July 1950, the exiled representatives of these parties established the Christian Democratic Union of Central Europe (CDUCE). Their political, journalistic and lobbying activity was focused mainly on fighting Communism, attacking the Soviet Union and liberating and democratising their countries. Political refugees in Latin America contributed to the establishment of the intercontinental network.

From the middle of the 1950s onwards the NEI lost its relevance. With the Coal and Steel Union and the foundation of the European Economic Community (EEC), practical cooperation among Christian Democrats gradually shifted in favour of the framework presented by the Common Assembly and the European Parliament. The organisation revitalised itself by changing its name to the European Union of Christian Democrats (EUCD) and revising the key aims of the organisation. The EUCD forged a closer relationship with the Parliamentary group of European Christian Democrats and the national member parties, and steadily grew more ambitious in its vision for Europe.

Direct elections to the European Parliament came about in 1979, and the need for a truly European party became evident. The formal establishment of the European People's Party (EPP) took place in 1976 in Luxembourg, with member parties from the following EEC countries: Belgium, Germany, France, Ireland, Italy, Luxembourg and the Netherlands. The platform was the result of considerable consensus and expressed a common intention to promote integration in the context of the European Community, leading to a political union equipped with federal and democratic institutions.

Once the EPP had been founded, a degree of pressure to establish formal links between Christian Democratic and Conservative forces was exerted by EUCD parties in countries that were not European Community members. Yet the EPP's strong insistence on the federal model of European integration led to the formation of the European Democratic Union (EDU), a broader pan-European organisation. Thus three parallel political organisations of Christian Democrats and Conservatives were now in place.

However, the EPP soon politically outweighed the EUCD, and the members who also belonged to the EPP concentrated more and more on their work in the latter group. The issue of merging the two organisations re-emerged when Spain and Portugal joined the European Community in 1986, but the revolutionary events which took place in Moscow and in other Eastern European capitals delayed the idea of a "big" EPP. Moreover, the EUCD's loose framework was better suited to the unclear political situation in the eastern countries; in fact, the organisation played an important role in supporting democratic progress and shaping the political landscape in the post-communist countries.

The political upheavals in 1989 meant that previous positions taken by the EPP had to be rethought and reformulated. The international context had been altered with the fall of the Berlin Wall and the end of the ideological conflict between East and West. And it was clear that the population of the German Democratic Republic wanted unification with the Federal Republic, as well as democracy. At the same time public opinion had shifted: the change enshrined in the Maastricht Treaty meant a political redefinition of Europe.

In April 1991, party and government leaders of the EPP decided that, while the party would be open to the British and Nordic Conservative parties, Christian Democracy would be preserved as the cornerstone of EPP identity. The EPP needed to integrate like-minded forces in order to achieve the majority needed to make ideas and concepts a reality. Although Greece's Nea Demokratia had already been admitted in 1983, in the early 1990s parties from Spain and the Nordic countries were included under the committed leadership of Wilfried Martens.

With the prospect of Central and Eastern European Countries joining the European Union (EU), the previous arguments supporting EUCD membership lost relevance – this led to the merger of the EUCD with the EPP in 1999. And since the EPP had accepted most European Conservative parties from the EU and beyond, the EDU also lost relevance, leading to its merger with the EPP in 2002. The development in the EPP reflected that of the EU itself; the inclusion of centre-right parties from accession countries in Central and Eastern Europe proved to be particularly successful. The new members brought a new dimension to the EPP and consolidated it as the pre-eminent European force of the centre-right.

By 2008 the EPP hosted 73 member parties in 38 EU and non-EU countries. Great Britain remains unrepresented, but the British Conservative Party has been allied with the EPP (since 1992) through its membership in the EPP-ED Group in the European Parliament.

Chronology

1925 First international congress of Christian (Catholic) people's parties (December, Paris). It was agreed to hold further meetings and to establish the "Secrétariat International des Partis Démocratiques d'Inspiration Chrétienne" (SIPDIC) in Paris; the Secretariat continued to exist until 1939. Parties from Belgium, Germany, Italy, France, the Netherlands, Luxembourg, Austria, Switzerland, Czechoslovakia, Hungary, Spain, Portugal and Lithuania were involved in its activities.

1946 Following the Second World War, cooperation between political parties at a European level was renewed. An initiative by the Swiss Christian Democrats led to the establishment of the "Nouvelles Equipes Internationales" (NEI).

1947 Constituent Congress of the NEI in Chaudfontaine (Belgium). The NEI pledged to cooperate actively in the reshaping of Europe at state, social and economic levels for peaceful coexistence and respect for human rights, liberty and social progress.

1948 As an active element of the European movement, the NEI participated in the preparations for the organisation of the famous "Congress of Europe" in The Hague.

1953 The Christian Democrat members of parliament of the six member states founded the first European Group of Christian Democrats within the parliamentary assembly of the European Coal and Steel Community.

1965 The NEI became the European Union of Christian Democrats (EUCD). Mariano Rumor (DC) was elected President. Leo Tindemans (CVP) was appointed Secretary-General.

1970 Establishment of a permanent conference within the EUCD of the Presidents and Secretaries-General of the Christian Democratic Parties of the Member States of the European Communities.

1972 Establishment of the "Political Committee" of the Christian Democratic Parties of the European Community with the aim of improving the coordination of European policy and cooperation.

1973–1974 Change of leadership in the EUCD: Kai-Uwe von Hassel (CDU) appointed President and Arnaldo Forlani (DC) Secretary General.

1975 Establishment of a "European Party" working group with the task of drawing up a draft Statute for a European union of parties. Wilfried Martens, Chairman of the CVP (Belgium), and Hans-August Lücker, Chairman of the Christian Democratic Group in the European Parliament were appointed as rapporteurs.

1976 The "Political Committee" unanimously approved the statutes of the European People's Party (EPP) on 8 July in Luxembourg. Leo Tindemans was elected President of the EPP. The following parties were the founding members: CDU and CSU (Germany), PSC and CVP (Belgium), CDS (France), Fine Gael (Ireland), DC (Italy), CSV (Luxembourg), KPV, CHU, and ARP (Netherlands).

1978 Congress I in Brussels adopted the political programme of the EPP.

1979 Congress II decided on the electoral platform for the first direct elections to the European Parliament. The EPP won 107 of the 419 seats in the elections.

1980 Congress III of the EPP in Cologne.

1981 As a result of the accession of Greece to the European Community, the number of seats in the European Parliament increased to 434. The EPP Group's share was 109 seats.

1982 Following the elections in Greece in June 1982, the number of MEPs in the EPP increased to 117. Congress IV of the EPP in Paris.

1983 Merger of the EUCD Secretariat (hitherto in Rome) and the EPP Secretariat in Brussels. Thomas Jansen appointed Secretary-General of the EPP and EUCD. Greece's Nea Demokratia joins the EPP.

1984 Congress V of the EPP in Rome adopted an action programme for the second electoral term of the European Parliament. The EPP won 110 seats at the second direct elections to the European Parliament.

1985 Piet Bukman (CDA) elected President. Thomas Jansen elected Secretary-General.

1986 Congress VI of the EPP at The Hague. As a result of the accession of Spain and Portugal, the number of seats in the European Parliament increased to a total of 518. The Portuguese CDS, the Spanish PDP (later renamed Democracia Cristiana), the Catalan UDC and the Basque PNV join the EPP. The size of the EPP Group increases by nine MEPs to 118.

1987 Jacques Santer elected President. Thomas Jansen re-elected Secretary General for a second term.

1988 EPP Congress VII in Luxembourg. Adoption of the working programme "On the People's Side".

1989 After the third direct election for the European Parliament of June, the Spanish MEPs of the Partido Popular join the EPP Group.

1990 Wilfried Martens elected President. Thomas Jansen re-elected Secretary-General for a third term. EPP Congress VIII in Dublin: adoption of the EPP programme for the European Union.

1991 The Spanish Partido Popular joins the EPP. Christian Democrat parties from Austria (ÖVP), Sweden (KDS) and Malta (PN) are admitted as associate members of the EPP.

1992 The British and Danish Conservative MEPs (together with some French UDF MEPs) join the EPP Group as allied members, bringing the total to 162 Members. The EPP Basic Programme is adopted at the EPP Congress IX in Athens.

1993 Nordic conservative parties are admitted to the EPP as permanent observers. The CDS of Portugal was expelled. Wilfried Martens re-elected President. Thomas Jansen re-elected Secretary-General for a fourth term. The EPP Congress X meeting in Brussels adopts the Action Programme "Europe 2000 – Unity in diversity" for the fourth parliamentary term of the European Parliament.

1994 Following the establishment of the Committee of the Regions (CoR) as a new institution of the European Union, the EPP Group was formed in the CoR with approximately 85 Members under the chairmanship of Jos Chabert, (CVP-B). The Christlichdemokratische Volkspartei (CVP) of Switzerland and the Democratic Rally (DISY) of Cyprus are admitted to the EPP as associate members. The EPP member-parties win 125 seats in the fourth European elections of June. The incorporation of like-minded MEPs (British and Danish conservatives and French liberals) bring the total number of EPP MEPs to 157. Klaus Welle elected Secretary-General of the EPP.

1995 The MEPs of the Kristdemokratiska Samhällspartiet and Moderata Samling (Sweden), Kansallinen Kokoomus (Finland) and the Österreichische Volkspartei (Austria) join the EPP Group. Kansallinen Kokoomus (Finland), Moderata Samlingspartiet and Kristdemokratiska Samhällspartiet (Sweden), Det Konservative Folkeparti (Denmark) and the Österreichische Volkspartei (Austria) become full members of the EPP. Høyre (Norway) accorded associate member status. Centro Cristiano Democratico (CCD) and Cristiani Democratici Uniti

(CDU) become full members of the EPP. Congress XI in Madrid "EPP - Force of the Union". Foundation of the European Senior Citizens Union (ESU).

1996 Foundation of the Small and Medium Enterprises Union of the EPP or SME-Union. In February seven parties in the prospective member-countries in Central and Eastern Europe are candidates for EPP observer status: KDU/CSL and ODS (Czech Republic), KDH and MKDM (Slovakia), KDNP and MDF (Hungary), PNȚCD (Romania).

1997 EPP Congress XII in Toulouse – "We are all Part of One World".

1999 Congress XIII in Brussels – 1999–2004 Action Programme – "On the Way to the Twenty-first Century". The congress re-elects President Wilfried Martens for another term and elects Alejandro Agag as Secretary-General, succeeding Klaus Welle who, in turn, is appointed Secretary-General of the EPP-ED Group in the European Parliament. Merger of the EUCD in the EPP formally concluded – EPP recognised as a regional organisation of the Christian Democrat International (CDI).

2000 EPP associate member status awarded to Tautas Partija (Latvia), SMK-MKP (Slovakia), FKGP and FIDESZ – MPP (Hungary). Merger of the Secretariat of the European Democrat Union (EDU, until then in Vienna) with the EPP Secretariat in Brussels.

2001 January: EPP Congress XIV in Berlin. Basic document approved: "A Union of Values". EPP associate member status awarded to MDF, Hungary. Full member status awarded to UDEUR, Italy and RPR, France.

2002 March: Political Bureau accepted, by acclamation, a proposal to replace Secretary-General Alejandro Agag by Antonio López Istúriz. EPP Convention Group established, following an initiative by President Martens. EPP associate member status awarded to EVP (Switzerland) and KDH (Slovakia). EPP observer member status awarded to SDKU of (Slovakia). October: Congress XV in Estoril, Portugal. Congress approves "A Constitution for a strong Europe". Merger of the EDU in the EPP formally concluded.

2003 In November, the EPP, as well as other European level parties, receive formal recognition by the European institutions, following the approval of "EU Regulation governing political parties at European level and the rules regarding their funding".

2004 February: XVI Congress in Brussels where the "Action Programme 2004–2009" for the June 2004 elections is approved. May: Following the accession of ten new EU Member States, all EPP associate member-parties from these countries become full members. June: In the first European elections, the EPP is victorious. The EPP-ED Group is, once again, the largest in the European Parliament with 268 MEPs. As a result of this victory, the EPP succeeds in the nomination of José Manuel Barroso by the European Council as the new President of the European Commission. September: HDZ of Croatia upgraded from observer to associate member. December: Observer membership granted to three Bosnian parties: the PDP, the SDA, and the HDZBiH.

2005 January: observer membership granted to the Turkish AK Party, and to Ukraine's "Our Ukraine" bloc. April: EPP President Wilfried Martens receives from the former German Chancellor Helmut Kohl the "Helmut-Kohl-Ehrennadel in Gold" distinction (Helmut-Kohl-honorary golden needle) for his exceptional services to Europe. June: EPP launches its first academic journal "European View". September: observer membership is granted to PD Romania.

2006 XVII Congress in Rome on 30 and 31 March – EPP adopts manifesto "The citizens and Europe: clear priorities for a better Europe". President Wilfried Martens and Secretary General Antonio López Istúriz are re-elected. In September, the EPP moved to its new headquarters, which better suited the Party's needs and objectives. April: US Senator John McCain addresses EPP event in Brussels on transatlantic relations. June: Two Belarusian parties are granted observer membership – the Belarusan Popular Front and the United Civil Party. November: Romanian PD is upgraded from observer to associate member.

2007 January: with the accession of Romania and Bulgaria, the respective parties are upgraded to full member status: PD, RMDSZ and PNȚCD (Romania) and DSB, UDF, BANU-PU and DP (Bulgaria). May: the Hungarian KDNP becomes a full member, and the VMRO-DPMNE from the Former Yugoslav Republic of Macedonia (FYROM), becomes an observer. September: HSS of Croatia upgraded from observer to associate member. Amendment of the "EU Regulation governing political parties at European level and the rules regarding their funding", mandates EPP and all European parties to campaign for the European elections. Amended Regulation also allows the creation of political foundations linked to political parties. Centre for European Studies (CES) established as the official think-tank/foundation of the EPP.

2008 February: GERB, Bulgaria becomes a full member, Batkivshchyna, Ukraine becomes an observer member. April: EPP recognised as a regional union by the International Democrat Union (IDU). September: UNM Georgia accepted as observer member.

- EPP Presidents
    - Tindemans, Leo 1976–1985
    - Bukman, Piet 1985–1987
    - Santer, Jaques 1987–1990
    - Martens, Wilfried 1990–

- Honorary Presidents
    - Niinistö, Sauli
    - Tindemans, Leo

EPP Congresses

1978 Congress I held in Brussels adopted the EPP's political program.

1979 Congress II also in Brussels, decided the electoral platform for the first direct elections to the European Parliament.

1980 Congress III in Cologne discussed the overall theme "The Christian Democrats in the Eighties – securing Freedom and Peace Completing Europe".

1982 Congress IV took place in Paris under the slogan "Preserve Peace – Create Peace – Unite Europe".

1984 Congress V in Rome formulated the EPP Action Program.

1986 Congress VI in The Hague focused on Economic Development and Environmental Problems.

1988 Congress VII in Luxembourg prepared for another European Election and approved the document "On the Side of the Citizens".

1990 Congress VIII in Dublin faced a new political landscape and published "A federal constitution for the European Union".

1992 Congress IX in Athens discussed, drafted, and approved the EPP "Basic Programme".

1993 Congress X was held in Brussels and adopted the action program, "Europe 2000: Unity in Diversity".

1995 Congress XI in Madrid was hosted under the thematic slogan "EPP-Force of the Union".

1997 Congress XII in Toulouse approved the document "We are all Part of One World".

1999 Congress XIII in Brussels outlined its new vision for Europe, "On the Way to the Twenty-first Century". EUCD formally merged with EPP.

2001 Congress XIV in Berlin revisited its roots, and adopted the working document "A Union of Values".

2002 Congress XV in Estoril prepared for the Constitution process, and declared that EU needed "A Constitution for a strong Europe". EDU formally merged with EPP.

2004 Congress XVI in Brussels where the main objective was to prepare for the European Elections. The "Action Program 2004–2009" was approved.

2006 Congress XVII in Rome approved the "Rome Manifesto" document.

Notes

Photos

Fig. 1 In August 1958 at the Yser pilgrimage, an annual day of remembrance for the Flemish soldiers who perished on the battlefields of the First World War, I addressed as a twenty-two-year-old student leader the crowd with a Flemish "State of the Union" speech.

Fig. 2 A far-reaching partnership developed between Jean-Luc Dehaene and me after he joined the youth wing of the Flemish Christian Democrats. His directness of speech and his non-conformity made him the right person for the job and he would succeed me as Prime Minister.

Fig. 3 During the party congress on 4 March 1972, I was elected President of the Flemish Christian Democrats, here with eminent members of our leadership: former Prime Minister Leo Tindemans, Vice President of my party Godelieve Devos, former Prime Minister Jean-Luc Dehaene and Jos Chabert, former Chairman of the Committee of the Regions.

Fig. 4 My experience of King Baudouin is one of a man of great integrity, high moral standards and an extreme sense of the importance of the State. Over the course of the years, I saw the King evolve from a moderate reformer to a convinced federalist [Photo Belga].

Fig. 5 Under the presidency of Charles-Ferdinand Nothomb and Frank Swaelen I addressed both Chamber and Senate, reading King Baudouin's letter declaring that his conscience would not allow him to sign and thereby to give assent to the law on abortion passed by Parliament.

Fig. 6. Together with my two liberal Vice Prime Ministers Jean Gol and Guy Verhofstadt. They were seasoned politicians with well-thought-out views who propagated a worthy form of socially tinted liberalism.

Fig. 7 "Would the United States really run the risk of a nuclear reprisal in order to defend the West in the event of a possible limited Soviet attack on Rotterdam or Hamburg?" German Chancellor Helmut Schmidt asked in October 1977. When I became Prime Minister in April 1979, he insisted that a number of other NATO countries besides West Germany should station the new intermediate-range ballistic missiles on their territory. He – quite rightly – wanted to share this nuclear responsibility with other European allies.

Fig. 8 14 January 1985 during my crucial visit to Washington on the cruise missiles. President Reagan was assisted by the Secretary of State George Schultz, the Secretary of Defense Casper Weinberger and National Security Adviser Robert McFarlane. Ronald Reagan had a good idea of what he wanted and set out towards his goal in a very straightforward manner. "No" meant "No" and "Yes" meant "Yes". On the Belgian side: Foreign Minister Leo Tindemans and my Chief of Staff Fons Verplaetse.

Fig. 9 US President George H. W. Bush maintained strong contacts with me, the EPP, and our centre-right political family as a whole. We remain until this day good friends.

Fig. 10 Although for diplomatic reasons I had called Mobutu "a friend of our country", he was never a friend of mine. At times, he was extremely arrogant and he gave me the impression that as a head of state and "Marshal" he felt himself to be superior to a simple Prime Minister.

Fig. 11 After a two year pilgrimage (1992–1994) to Angola, Burundi, Ethiopia, Kenya, Namibia, Rwanda, Sudan, Somalia and South Africa with the noble teams of Médecins sans Frontières, I asked "Africa is dying, do we care?".

Fig. 12 At the European Summit preparing the Single Act with President François Mitterrand and my colleagues Prime Ministers Margaret Thatcher, Bettino Craxi, Pierre Werner and the German Foreign Minister Hans-Dietrich Genscher.

Fig. 13 I always got on extremely well with Helmut Kohl. Our convictions on Europe are identical. I went to all the CDU congresses and have continued this tradition right up to the present day. We were members of the European Council for almost ten years. We always worked together closely and with mutual trust, not only on the Council but also within the EPP. Once you gain his trust, his support is unconditional.

Fig. 14 From the start something had clicked between Ruud Lubbers and me. I still regard him as a good friend. Like so many Dutch politicians, he was closely attuned to London. "That is my role", he told me. He was probably the only European leader who maintained good contacts with Margaret Thatcher.

Fig. 15 With José María Aznar I shared an increasing degree of mutual trust. Our conversations have always been open and honest. I have never hesitated to support him politically, in contrast to other heads of government within the EPP. I continued to pursue this line and attended all of the Partido Popular's election campaigns during their period of growth.

Fig. 16 Jacques Delors was in essence a Christian Democrat, even though he belonged to the Socialist family. He is a practising Catholic and knows the classic tenets of Christian Democracy – personalism, federalism, subsidiarity – better than many politicians who call themselves Christian Democrats. Moreover, he is exceptionally pro-European.

Fig. 17 John Bruton with his successor as Fine Gael leader and EPP Vice President Enda Kenny; in the middle, leading UMP politician and EPP Vice President Michel Barnier. John Bruton was one of the most European-minded Prime Ministers ever and also a very active and dynamic Vice-President of the EPP. What he fought for as chair of the Athens Group – the preservation of the Christian Democratic roots of an ever-broadening EPP – is in complete agreement with my deepest convictions.

Fig. 18 Prime Minister Giulio Andreotti with the Italian Christian Democratic leaders Forlani and Piccoli. His contribution to the success of the IGC on the internal market and the Euro is indisputable. Andreotti is extremely intelligent, a convinced European, inventive, mysterious and intriguing. In spite of his advanced age, he is still active in Italian politics. He is a true survivor. On his right, Nea Demokratia leader Constantine Mitsotakis, Prime Minister during the re-launching of Europe in the eighties and host of the EPP Congress in Athens in November 1992.

Fig. 19 With Hans-Gert Pöttering and Ria Oomen-Ruijten during the 1994–1999 legislative term of the European Parliament. They were my Vice-Presidents of the EPP Group in full expansion. Hans-Gert Pöttering was elected Chairman of the EPP-ED Group in 1999 and President of the European Parliament in January 2007.

Fig. 20 Among the spiritual leaders I received in the European Parliament was the highest teacher of Tibetan Buddhism, the Dalai Lama, His Holiness Tenzin Gyatso.

Fig. 21 In March 1997, the EPP Group attended an audience with Pope John Paul II at the Vatican for the fortieth anniversary of the Treaty of Rome. The Pope was still in good health at the time and in our conversations it was the contention, now that the Berlin Wall had fallen, that "the hour of the Christian Democrats had arrived".

Fig. 22 Joseph Daul was elected Chairman of the EPP-ED Group in January 2007. We created an open and strong relationship between the Party and the Group.

Fig. 23 Jacques Santer presiding at the seventh EPP Congress. On 15 July 1994 Helmut Kohl put him forward as the new Commission President during a special meeting of the European Council. Given his many years of experience as Prime Minister, the choice was obvious. He had also been President of the EPP for four years and shared our convictions on Europe. We placed much hope in his presidency, which, unfortunately, would not end favourably. Because of events beyond his control he was judged negatively in retrospect, which was unfair.

Fig. 24 The political vacuum created by the implosion of the Democrazia Christiana in Italy needed to be filled and there was really only one candidate: Silvio Berlusconi and his Forza Italia.

Fig. 25 Prime Minister Berlusconi invited in Sardinia his EPP colleagues Prime Ministers José-Maria Aznar, Jan-Peter Balkenende, José-Manuel Barosso, Jean-Claude Juncker, Jean-Pierre Raffarin, and also Hans-Gert Pöttering and myself.

Fig. 26 In 2004 José Manuel Barroso had the courage to give up his premiership of the Portuguese government in favour of the presidency of the European Commission, "the most difficult job in the world", according to the title of a French TV documentary.

Fig. 27 Jean-Claude Juncker's commitment to Europe is deeply rooted in the traditions of the Christian Democrats. Prime Minister of Luxembourg since 1995, he is a strong voice at EPP summits and in the European Council. His prestige is also the result of his expertise and efforts as chair of the Euro Group.

Fig. 28 Former Prime Minister Jean-Pierre Raffarin was the first French Prime Minister ever to participate in the EPP summits.

Fig. 29 Brok's Group inside the European Convention was perhaps one of the most energetic and vital we have ever known. The most striking persons present here in the South of France are former German Chancellor Helmut Kohl, the President of the Convention Valéry Giscard d'Estaing and former Prime Minister of Denmark Poul Schlüter.

Fig. 30 In 2004 the Netherlands rejected the European Constitution in a referendum. Today a change is taking place. The question is whether the Dutch political class will still be capable of following the European traditions of its predecessors. All eyes are fixed on the Prime Minister, Jan Peter Balkenende, to restore this continuity. I am convinced that he will succeed.

Fig. 31 The German Christian Democrats form the backbone of the EPP and the European Union. That was the case in the time of Helmut Kohl and turned out to be so again during the German EU Presidency in the first half of 2007. Angela Merkel performed her challenging and sensitive task magnificently. Thanks to her personal commitment, dedication and efforts, considerable progress has been made in three important domains: the EU multi-year budget, the fight against global warming and the salvaging of the European Constitution through the Lisbon Treaty.

Fig. 32 The French President Nicolas Sarkozy with Prime Minister José María Aznar and CDI President Pier Ferdinando Casini. With the UMP, the EPP has gained a steady partner in France and our lines of communication reach to the very top of French politics. Nicolas Sarkozy is strongly committed to the EPP: at a crucial moment in 1999, he pushed his MEPs to become members of our Group. His participation at our congress and summit meetings has given them a particular brilliance. He was a strong President of the European Council during the crucial second half of 2008.

Fig. 33 Since March 2004, Nea Demokratia is once again in power and Greece has a pro-European Prime Minister, Kostas Karamanlis, a two-term EPP Vice-President. In 1999, Karamanlis launched the Western Balkan Democracy Initiative in an effort to identify and propose potential partners for the EPP. Today, most of these parties – like Sali Berisha's Democratic Party – are observer or associate members of the EPP. In Albania Berisha supported the student strike in December 1990, which turned political and obliged the government to allow the creation of new parties. The same month the Democratic Party was founded by students and intellectuals and in 1991, Berisha was elected its leader. After the first free elections, he became President of Albania in April 1992.

Fig. 34 In November 2007 Donald Tusk became Prime Minister of Poland and formed an EPP government with the PSL leader Waldemar Pawlak. He was one of the founders of the Civic Platform and from June 2003 he took over as party leader. Donald Tusk represents a new Poland that is focused on the important values of Western civilisation. He is our great hope for a real European Poland. From 1998 until 2006 Mikuláš Dzurinda has changed the image of Slovakia entirely through political and economic reforms and membership in the European Union and in NATO and secured harmonious cooperation with the minorities of his country.

Fig. 35 The low profile adopted by Wolfgang Schüssel had helped Austria to emerge from the *Cordon sanitaire* imposed by the European Council. He was proven right in his strategy of challenging the extreme right via a coalition. He did not make compromises and he took no initiative that was not within international and European rules of law.

Fig. 36 Swedish Prime Minister Fredrik Reinfeldt was an active leader in the EPP youth movement. He belongs to the first generation of leaders maturing in a European political context.

Fig. 37 Jyrki Katainen is the leader of our member party in Finland, Kansallinen Kokoomus. In 2006 he was elected Vice-President of the EPP; in March 2007, his party did very well in the general elections and he was appointed Deputy Prime Minister and Minister of Finance. He also leads an informal gathering of EPP Ecofin Ministers. On his right Mart Laar: his Pro Patria Union came into government and as Prime Minister he managed to lead Estonia through the lightning economic reforms that won praise and ultimately laid the groundwork for rapid economic growth and acceptance into the European Union. Per Stig Møller is Denmark's successful Foreign Minister from the Conservative People's Party.

Fig. 38 Viktor Orbán was a founding member of Fidesz (Alliance of Young Democrats), who were persecuted by the Communist Party. The movement became a major force in many areas of modern Hungarian history. On 16 June 1989, Orbán gave a speech at Heroes' Square on the occasion of the reburial of Imre Nagy and other national martyrs, in which he demanded free elections and the withdrawal of the Soviet troops. In the summer of 1989 he took part in the Roundtable negotiations and in 1998 he became Prime Minister. Fidesz was transformed from a radical student movement into a moderate, centre-right people's party. On my left side former French Prime Minister Alain Juppé. With his drive as RPR party leader and founder of the UMP (and with the support of President Jacques Chirac) neo-Gaullists managed to shake-off their Eurosceptics and joined the EPP. In the middle of the photo Prime Minister (at the time EPP Vice President) Kostas Karamanlis and EPP Secretary-General Antonio López-Istúriz.

Fig. 39 Ivo Sanader has emerged as the leading political figure of his country. When he took over the leadership of the Croatian Democratic Union (HDZ), he managed to transform it from a nationalist party to a modern, pro-European centre-right party that wants to develop good neighbourly relations with all of its former rivals. Sanader led the HDZ to victory in the 2003 and 2007 parliamentary elections and is poised to make history as the Prime Minister of the twenty-eighth EU Member State.

Fig. 40 After the fall of Slobodan Milošević as President of Yugoslavia in October 2000 we were looking for a democratic alternative. I visited Zoran Djindjić (pictured), the leader of the Democratic Party. He had studied in Germany and was friends with Willy Brandt. His party joined the Socialist International. We sought contact with Vojislav Koštunica; he was both a Western liberal who believed in a free press, an independent judiciary and multi-party democracy, and at the same time a Serb nationalist often critical of the Western alliances but with no connection to the old Communist Party.

Fig. 41 In his letter of 8 September 2004, Prime Minister Erdoğan stated: "In the preamble to the statutes of the EPP there is a reference to the Christian view of mankind and the Christian Democratic concept of society. The AKP will abide by this provision to the extent allowed by its own statute and in conformity with the Constitutional Treaty of the EU". I put the following question to him: "Why does not AKP call itself Muslim Democrats as we in the EPP call ourselves Christian Democrats?" His answer was clear: "We can never use our religion in the name of the party because people, movements and parties make mistakes and this would have a devastating effect on religion". Instead, the AKP officially defines itself as Conservative Democrat.

Fig. 42 The Orange parties obtained the majority in Ukraine's parliament. The President, Viktor Yushchenko and the Prime Minister, Yulia Tymoshenko, participated at the EPP Summit in Lisbon on 18 October 2007. They committed themselves to form an "Orange" government and thanked our leaders for their support as well as for the recognition of the Holodomor, or forced famine. In 1932 and 1933 millions of innocent victims died of hunger in Ukraine because of the orders of Stalin's regime.

Fig. 43 Georgia's President Mikheil (Misha) Saakashvili and his United National Movement (UNM) had in May 2007 applied for EPP membership. When I asked in a meeting with NGO representatives if they considered this remote country to be a European country, one of them responded, "Georgia cannot afford not to be a European country. If Georgia ceases to be European then it will cease to be a country".

Fig. 44 Senator John McCain – also Chairman of the International Republican Institute (IRI) – opened the door for the close cooperation between the EPP and the IRI. McCain is convinced of the importance of strong EU-US relations.

Fig. 45 The EPP Summit in Berlin at the fiftieth anniversary of the Treaty of Rome.

GPSR Compliance

The European Union's (EU) General Product Safety Regulation (GPSR) is a set of rules that requires consumer products to be safe and our obligations to ensure this.

If you have any concerns about our products, you can contact us on

ProductSafety@springernature.com

In case Publisher is established outside the EU, the EU authorized representative is:

Springer Nature Customer Service Center GmbH
Europaplatz 3
69115 Heidelberg, Germany

www.ingramcontent.com/pod-product-compliance
Lightning Source LLC
LaVergne TN
LVHW011001250326
834688LV00003B/47